*Roadside Bedrock Geology along Route 129 and 101*
*(from Thessalon to Potholes Provincial Nature Reserve, Ontario)*
*Steven D.J. Baumann*
*©2019*
*Midwest Institute of Geosciences and Engineering*
*(inside cover)*

I would like to personally thank the following individuals who have made this book possible by either joining me in the field, selecting the stops, editing, or just by encouraging my love for rocks. I dedicate this book to all of you!

David H. Malone

Sarah M. Hall

Elisa J. Piispa

Sandra K. Dylka

Veronica George

Dave Johnson (new logo designer)

---

Cover Photos taken by Steven D.J. Baumann on November 24, 2018. Background photo is of the quartzolite vein at Stop 19. The waterfall is Chute Falls Provincial Park at Stops 8 and 9.

# ROADSIDE BEDROCK GEOLOGY ALONG ROUTE 129 AND 101

## From Thessalon to Potholes Provincial Nature Reserve, Ontario

### STEVEN D.J. BAUMANN

**Midwest Institute of Geosciences and Engineering**

© 2019

www.mige-web.org

| Table of Contents: | Stop Number: | Page: |
|---|---|---|
| Introduction | | 2 |
| Geologic Terminology | | 4 |
| The Rock Cycle | | 10 |
| Cross Section of the Plate Tectonics Cycle | | 11 |
| Igneous Parent Rocks and Metamorphic Equivalents | | 12 |
| Sedimentary Parent Rocks and Metamorphic Equivalents | | 13 |
| Classification Chart of Coarse Grained Igneous Rocks | | 14 |
| Applicable Geologic Timescale | | 15 |
| Huronian Supergroup | | 16 |
| Major Types of Faults | | 17 |
| Large Scale Geologic Structures | | 18 |
| How to Read a Geologic Map | | 19 |
| Bowen's Reaction Series | | 20 |
| Outcrop Stops: Location Maps | | 22 |
| Maps to Current Roadside Books of Ontario | | 24 |
| Route 129 and Trans-Canada 17 Outcrop | 129-1 | 26 |
| Route 129 and Station Road Northeast Outcrop | 129-2 | 28 |
| Route 129 North of Yates Lane Outcrop | 129-3 | 30 |
| SE Side Route 129 East of Appleby Lake Outcrop | 129-4 | 32 |
| North Side of Tunnel Lake Outcrop | 129-5 | 34 |
| Route 129 east side of Lafoe Creek south Outcrop | 129-6 | 36 |
| Route 129 east side of Mississagi River Outcrop | 129-7 | 38 |
| Route 129 Aubery Park entrance Outcrop | 129-8 | 40 |
| Route 129 opposite Flame Lake north Outcrop | 129-9 | 42 |
| Route 129 between Nemi and Pike Lakes Outcrop | 129-10 | 44 |
| Route 101 1.4 mile west of Route 129 Outcrop | 101-12 | 46 |
| Route 101 west of kilometer marker 123 Outcrop | 101-11 | 48 |
| Route 101 kilometer marker 119 Outcrops | 101-10 | 50 |
| Route 101 kilometer marker 94 east Outcrop | 101-9 | 52 |
| Route 101 just north of kilometer marker 83 Outcrop | 101-8 | 54 |
| Route 101 on Quill-Nadjiwon Township line Outcrop | 101-7 | 56 |
| References | | 58 |

# Introduction

This book explores only a handful of the many outcrops along Ontario Routes 129 and 101 from Thessalon to "Potholes Provincial Nature Reserve", Ontario. The stretch exposes mostly the crystalline igneous Archean rocks and the meta-sedimentary Paleoproterozoic Cobalt and Group of rocks. These rocks are about 3.2 to 2.2 billion years old and show some significant events in Earth's history. The Cobalt Group records the first time free oxygen ($O_2$) appears in the atmosphere. This book is the completion of a loop. Volume 4 in this series titled, is the western part of the loop. See page 24 for maps of the Ontario Guidebooks.

Perhaps the biggest pet peeve I have about roadside geology books is how they are organized. Traditionally you start out at one point. Then you go X-distance until you see an old yellow barn (or other landmark). Then you turn left and travel Y-distance to an outcrop. For one thing, if you pass it you're in trouble and you may have to double back. This can be a huge time waste if the outcrops are dozens of miles apart. It restricts you to a certain set order of visits, because it goes from point A to B, from B to C, then from C to D, etc. It is a poor way to navigate a roadside geology book. Landmarks are not stagnant. Barns fall down. Dirt roads become paved. Road signs change. Rob's Agates and Beer goes out of business and gets sold, and so on. This method was fine for the 20th century, but is archaic in an era of smartphones and GPS. Fortunately, technology does advance. Almost everyone has a global positioning system (GPS) receiver either in their vehicle or on their smartphone. It's not common knowledge, but if you directly type in latitude and longitude as below into any smartphone and most vehicle GPS systems, it will take you to the exact spot. I demonstrate this in the below two photos.

**PHOTO A** **PHOTO B**

 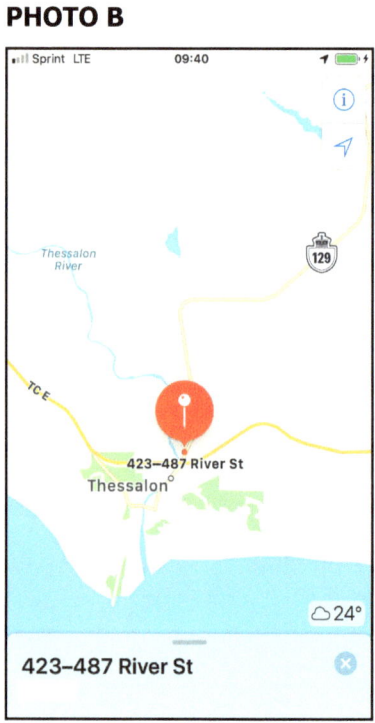

Here I type in the exact coordinates for my iPhone. It is for the stop (129-1). **PHOTO A:** You type in the latitude 46.26678 (no minus because you are in the northern hemisphere). Then the longitude -83.54835 (you need the minus because you're in the western hemisphere). **PHOTO B:** Then hit "search" or "enter" and it will drop you right at the spot. I record coordinates to within a several feet (a couple of meters).

Using GPS is a big help. You can do your outcrop visits in any order that you want! You just have to remember to type the decimal latitude first then the decimal longitude with a minus in front. All the stop GPS's in this book are latitude first then longitude. Copy it into your device directly.

The book is laid out so when you open it to a stop you see all the information. The basic info and geologic map on the left and the photos on the right. This way you can see all the information for a site without flipping pages.

All the outcrops in this book are on public land or in the public right-of-way. There were other outcrops that I wanted to add but didn't because you can't safely park, or something similar had already been visited. Also, I generally included taller outcrops so you can visit in the winter.

We visited this area July 2-3, 2017. All the photos are from those dates, and by Steven Baumann or Sarah Hall. The Inset cover photo is by Sarah Hall. All are road outcrops the book ends at Potholes Provincial Nature Reserve. Potholes is not included herein as it was included in Volume 4, Roadside Geology of the Algoma District, Ontario: From the Goulias River, to Wawa, to Potholes Nature Reserve. Potholes is technically closed in the fall and winter. Dates of operation may vary slightly from year to year. In 2019 it opened on June 14 and closed on September 2. Camping is not allowed.

The website for "Potholes Provincial Nature Reserve" is: ontarioparks.com/park/potholes

There are a couple of other parks that were not visited while we were on this route, but are right off the roads. The first is "Aubrey Falls Provincial Park". **See page 40, (Stop 129-8)**. We did not stop at this one because we weren't sure if the weather was going to hold. We got luck and it did, but you can never know. There were thick clouds to the north that ended up dissipating.

Along the Route 101 part of this guidebook (east of Potholes), we passed through the large "The Shoals Provincial Park. However, it was closed when we were there in the summer of 2017. It is open now. **See page 52 for website, (Stop 101-9)**.

This book only deals with the Precambrian rocks (the Archean and Proterozoic mostly). The reason being is that only the Precambrian forms the local bedrock. There are no Quaternary glacial stops. Glacial outcrops are rare and where they do exist, they tend to be low lying and get snow covered fast.

As you travel from south to north, the rocks get progressively older. Eventually you leave the meta-sedimentary and igneous Proterozoic rocks of the and enter the mostly plutonic and metamorphic Archean rocks.

I hope you enjoy this book and have many delightful and exciting discoveries!

# Geologic Terminology

Geologists, like most scientists, like to name things and come up with difficult to understand terms. The result can lead the average user to lose interest in the literature. I attempt to keep the technical jargon to a minimum. However, sometimes it is necessary. The names of the formations, for example, are much easier to use than to describe them each and every time. When I describe the different grades of metamorphism, I avoid terms like "greenschist facies" and "amphibolite facies". Instead I use "low grade" or "high grade". It is important to have a decent understanding of the metamorphic process in the area. This section is dedicated to the explanation of terms considered unavoidable.

**Basalt, Gabbro, Diabase, Andesite, and Rhyolite**

Basalt, gabbro, andesite, diabase, and rhyolite are all fine grained igneous rocks that are commonly associated with rifts and form directly from magma and lava. Basalt contains less than 30% quartz and feldspar and is usually dark colored. Basalt makes up most of the ocean floor. Gabbro is a coarse grained version of basalt and usually forms deep within the Earth. Andesite is similar to basalt except it tends to contain less iron minerals, like olivine. The two can be difficult to separate in the field. In the field the term diabase is used as a generically to describe a dark colored, fine grained igneous rock of variable or indeterminate composition. Rhyolite is chemically the same as granite, except it is fine grained. Most of these rocks are deposited on the surface, unless they are in dikes and sills. Since they usually form on the surface they are referred to as extrusive igneous rocks. Large intrusions that form underground tend to be coarse grained because they take longer to cool forming larger crystals, thus forming granitic rocks.

**Bed**

Beds are the basic divisions of layered rocks. Beds are defined by well marked surfaces called bedding planes and are where strike and dip is obtained to assess the orientation of a formation in the field. Beds can be divided into thin, medium, or thick. Beds >39.4" (100cm) are referred to as "massive" and beds <0.39" (1cm) are called laminations. Well defined continuous stacks of distinct or repeated beds are called "bed sets".

**Breccia and Clasts**

Breccia is used when a rock body contains angular blocks within a matrix or groundmass of different composition. Although usually reserved for igneous rocks, the term can be applied to any angular clastic rock, like sedimentary alluvial fans or debris flows. Rounded blocks are usually referred to as clasts and form the basis of conglomerates in sedimentary rocks. Clasts are also any detritus particle of any size. This can include breccia. The term "clastic rock" is used generically to refer to any rock composed of clasts. The term "clastic" excludes all rocks of biological or chemical origin (such as evaporites, carbonates, and coal).

**Chert**

Chert is usually a light colored rock made entirely of microscopic quartz. It is commonly formed from biological activity in the ocean. Most Precambrian chert formed differently, in that it is made of microscopic clastic quartz particles derived from wind blown deposits such as silt.

## Country or Host Rock

The rock surrounding the body of an igneous intrusion. It is the rock that has been intruded, it "hosts" the intrusion. Country rock is usually used when describing the emplacement of the intrusion, in a general or regional application. Host rock is usually used when describing something in the field at the outcrop scale.

## Cumulate

An accumulation of coarse crystals due to settling or floating of minerals as a magma slowly solidifies.

## Detrital

Detrital (or detritus) or fragments of rock of any size made of lithic (rock) fragments, usually broken down by mechanical weathering.

## Diamicton

Diamicton is a relatively new term. It is essentially equivalent to glacial till. However, diamicton does not infer an origin. It is usually used to describe deposits left directly by a glacier, yet it can describe any matrix supported conglomerate. The rock form of diamicton is still referred to as tillite, although diamictite is gaining acceptance.

## Faults

Faults are a break in the Earth that forms a planar surface in which movement has occurred. The vast majority of faults have been inactive for millions (or billions) of years. Most earthquakes occur along active faults. The movement can be up, down, sideways, or a combination of any. Most faults are caused by tectonics. As rock becomes compressed or extends it can either fold (ductile deformation) or break (brittle deformation). If it breaks, a fault is created, energy is released, and an earthquake occurs. A graphic representation of the main types of faults is shown on page 17. Some faults have been so compressed that the plane on which the movement occurred is no longer easily visible. These types of faults are called shear zones.

## Granite, Tonalite, Syenite, and Diorite

Granite, tonalite, syenite, and diorite are all coarse grained igneous rocks that are commonly associated with deep magma chambers. Along subduction zones and less commonly rifts. The difference between all these rocks is the relative amounts of quartz, plagioclase, and alkali-feldspar in the rocks. Typically these rocks form deep underground and will rarely exist as dikes and sills. The term "granitic rock" is a generic textural term that has nothing to do with the actual rock type. It is just used to describe a non-foliated part of a gneiss or for any rock that looks like granite at a quick glance or any rock that expresses a granite like texture.

## Joints and Fractures

There are many types of breaks that occur naturally in rock where no movement has occurred. As a rock is exposed to the surface through erosion, it will often "crack". Cracks can also form in the subsurface. These cracks are called fractures or joints. Although the two are different, herein they are treated as the same feature to avoid getting too technical. Rocks can also form these features from tectonic stresses. They may also form if a melting glacier retreats. As the weight of the ice is removed the rocks near the surface will often form fractures. Sometimes the fractures become filled with other minerals and close.

## Matrix, Cement, and Groundmass

The matrix is what binds the particles of rock together (in sandstones) or the fine grained portion of any sedimentary rock containing significant coarse grains. In sandstones the matrix is often the same as the cement (what holds the grains together). The most common minerals that make up the matrix or cement in a rock are calcite and quartz, although other minerals and small grains can serve as matrix. Cement is not used for igneous and metamorphic rock. The term groundmass is often used interchangeably for matrix, but this is not entirely correct (even I have been known to swap the two). Groundmass usually pertains to fine grained igneous and metamorphic rocks and is generally only used if the rock is porphyritic or has isolated large crystals.

## Metamorphism

Rocks that have been metamorphosed have undergone deep burial and have been altered through heat and pressure but not enough to melt the entire rock. The greater the heat and pressure, the more likely the rock is to become altered and form different minerals from existing ones. The vast majority of the Paleoproterozoic rocks in along the north shores of Lake Superior and Lake Huron are metamorphic but they have undergone low grade metamorphism. This means that they were buried deep enough to become metamorphosed but not enough to have altered their mineral make up. The Archean rocks were buried far deeper and longer than the Paleoproterozoic rocks. They tend to have higher grades of metamorphism. Some high grade metamorphism has occurred near the many extinct faults in the area. Metamorphism also occurs in close proximity of dikes. Metamorphism that occurs locally near faults and intrusions is often called contact metamorphism.

## Migmatite

A rock consisting of a macroscopic, discernable metamorphic component within an igneous component or vise versa. They form under maximum metamorphic temperatures and pressures that are so great, part but not all of the rock melts before recrystallizing. Migmatites are massive bodies. Igneous dikes within metamorphic rocks are not migmatite because dikes can intrude any host rock at any depth. High pressure and temperature of the host rock is not needed for dikes to intrude it.

## Parent Rocks

Parent rocks are the sedimentary or igneous rocks that exist before undergoing metamorphism.

**Phenocryst**

Phenocrysts are numerous large and obvious crystals within the groundmass of a rock, and are usually a different color than the groundmass. They are significantly larger than the next discernable crystals. There really is no standard for what constitutes "significantly larger". If I can see the crystals from 8-10 feet (a couple of meters) away, then I consider the rock porphyritic (the adjective used to describe a rock with phenocrysts).

**Quaternary**

The Quaternary is the period of geologic time from 2.4 million years ago to the present. It encompasses all of the modern ice ages. Quaternary deposits are common in the area but rarely exposed. They do commonly leave scour marks and striations on the Precambrian rocks.

**Rifts, Passive Margins, and Wilson Cycles**

Where the earth's crust begins to rift apart (split) due to upwelling magma chambers from deep in the Earth. If the rift persists eventually a continent will split and a new ocean will form between them as they move further and further part. Not all rifts form new ocean crust. The Mid-continental rift (under Lake Superior) is a failed rift that formed around 1.1 billion years ago. The rift that led to the deposition of the Huronian Supergroup was a successful rift, like the one that created the Atlantic Ocean. The initial and early stage of rift formation is due to active volcanism. Rifting usually occurs on the continents, but can occur at the bottom of the Ocean, like the modern "East Pacific Rise". As the rift matures, volcanism on the continent stops and a passive margin forms. The passive margin becomes a place for thick sequences of marine sediments to accumulate until another continent or island arc merges with the passive margin. The force that splits and moves continents around the globe is called Plate Tectonics. A Wilson Cycle (named after the Canadian geophysicist John Tuzo Wilson, 1908-1993) is the name given to the opening of an ocean basin and its closing, including all the steps in between. This includes rifting, passive margin development, subduction along the passive margin, and the accretion of island arcs or continents as the ocean plate shrinks due to subduction.

**Sandstone and Quartzite**

Sandstone is the rock version of sand. Sand is sediment and sandstone is a sedimentary rock. It consists of grains of sand ranging from 0.2 micrometers (about as small as the human eye can see) to 2 millimeter in diameter (roughly the width of two fingerprint lines). Sandstone is derived from other rocks that have been weathered from larger rock bodies. Sand can come from any rock. Feldspar rich sands are called arkoses. Quartz rich sands are what people typically think of when we say sand. Sand that consists of fragments other than feldspar or quartz are referred to as lithic sands. Sands with a lot of fines (silt or clay) are called wackes. Sand that is made up of mostly sand-sized particles are called arenites.

Quartzite is the metamorphic version of sandstone. Quartz rich quartzite is just called "quartzite". Arkosic quartzite is called "proto-quartzite". Quartzite rich in lithic fragments is called "immature quartzite". Silty or argillite rich quartzite are referred to as impure quartzite, or impure proto-quartzite, or impure immature quartzite.

**Shale, Mudstone, Argillite, and Slate**

Shale is a fine grained rock (smaller than sand) mostly made of clay. Mudstone is a sandy or silty shale. Argillite and slate are metamorphic versions of shale and mudstone.

**Strike and Dip**

This is a the basic measurement of rocks in the field and is the key measurement in the field. Strike is simply the line between two points of equal elevation always measured from north. These points can be hundreds of feet, several meters, or a couple of inches or centimeters apart. Dip is always perpendicular to strike in the down-slope direction given in degrees from the horizontal. When you combine the strike with the dip you get the true orientation or trend of a rock unit. The measurement is obtained with a Brunton compass and give in quadrant format always oriented to the north. A strike of 280° would be written as N80W. Dip, by definition has to be either NE or SW if your strike is NW because dip is always at a 90° angle down-slope to strike. If your strike is NE than dip has to be either NW or SE (see below for examples). Strike and dip can be used to record the orientation of any planar surface. This is usually bedding planes, and in this book it will be bedding unless otherwise noted. Strike and dip can also be used to measure fault planes, joint faces, fracture planes, and dike orientation.

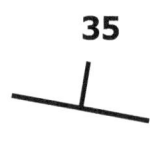

To the left is a graphic depiction of strike and dip. The long line without a number attached to it is the strike (in this case it trends N80W). The long line is oriented to the strike but contains no numbers. The smaller line indicates the dip with the degree of dip from the horizontal indicated by a number. In this case the rock dips 35° to the NE, and the degree (°) symbol is dropped. We would write out this trend or orientation in this example as **N80W 35NE**.

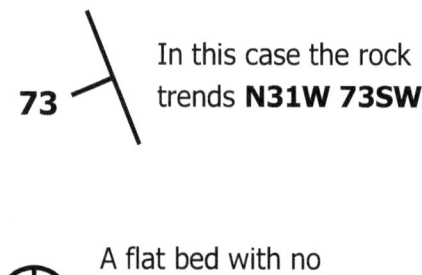

In this case the rock trends **N31W 73SW**.

In this case the rock trends **N45E 9SE**.

A flat bed with no dip is depicted by a cross in a circle and contains no numbers.

This symbol is used for vertical trends. In this case the rock trends **N57E 90**. There is no NE or SE attached to the 90 in a vertical orientation. No number is attached to the symbol.

## Subduction Zones

Subduction zones form where two plates collide and one subducts under the other, retuning it to the mantle (the layer in the Earth beneath the crust and lithosphere). When two continents collide only the ocean crust is subducted into the mantle. Continents are lighter and more buoyant so they do not get subducted. As ocean crust slides under a continent on its way back into the mantle, volcanoes form. Some will form on the continent. This is currently happening along the west coast of North and South America from Alaska all the way to Argentina. Sometimes volcanoes will form on the ocean crust and create island arcs. Japan is an island arc. Eventually island arcs are pushed against a continent and become a part of it. This is essentially what occurred during the Penokean Orogeny in Wisconsin, the Upper Peninsula, Minnesota, and Ontario about 1.76 to 1.84 billion years ago.

## Tonalite-trondhjemite-granodiorite (TTG)

Tonalite and granodiorite appear on the QAPF diagram, so they will not be elaborated upon here. Trondhjemite is essentially tonalite where all the plagioclase is oligoclase with little to no anorthite. TTG's usually exist as massive plutons in Hadean up through Neoarchean igneous complexes. Prior to the Mesoarchean TTG's were common globally and true granite was rare to non existent. By the beginning of the Paleoproterozoic TTG's became exceedingly rarer as Plate Tectonics took over and granite plutons became the normal. TTG's are still deposited today but are restricted to ophiolites and volcanic arc batholiths. No one knows why the Earth now favors granites over TTG's . It likely has something to do with the Plate Tectonics cycle which was not in operation during the Hadean and most of the Archean.

## Xenoliths

Large foreign inclusions in igneous rocks of older igneous, sedimentary, or metamorphic rock that is brought up from depth with rising magma but does not get melted. They are usually composed of the country rock but can be mantle fragments. They can be angular or rounded.

# The Rock Cycle

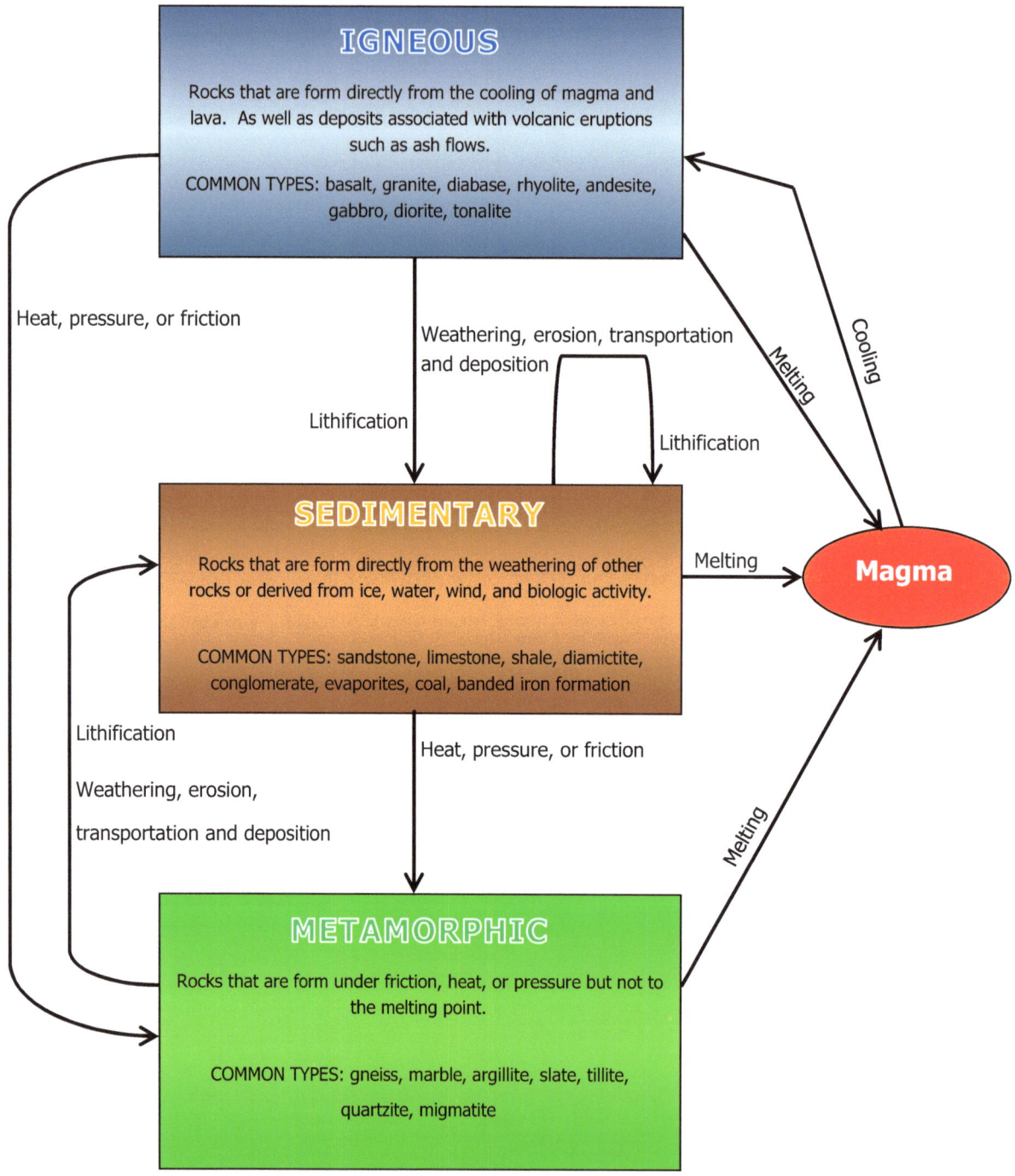

The rock cycle is the process of how one rock type becomes another. The above chart illustrates how the process unfolds. Not all rocks will become a different type, some will repeat the process again and again. The process can end anywhere on the chart. Plate Tectonics and the water cycle are the main driving mechanisms behind the rock cycle.

# Cross Section of the Plate Tectonics Cycle

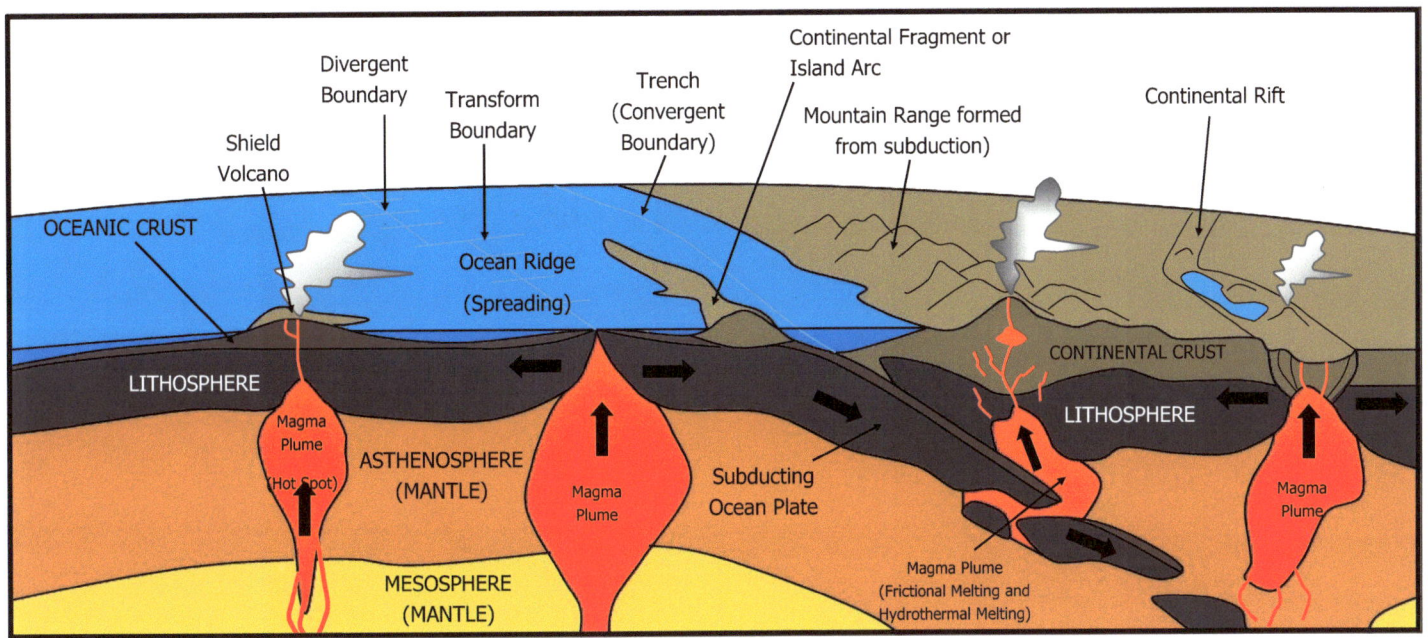

## The Earth's Interior

Plate Tectonics is main mechanism of how Earth loses its internal heat. Modern Plate Tectonics has been in effect for about 2.5 billion years (Earth is 4.6 billion years old) and will likely continue for a billion more. Prior to Plate Tectonics Earth likely lost its internal heat in a similar way that Venus does with a transitional period from about 2.5 to 3.2 billion years ago.

The inner core is a solid ball of iron and nickel, and its temperature is about 10,500°F (hotter than the surface of the sun). The inner core loses heat through conduction. The exact nature of the inner core is still largely unknown.

The outer core is liquid iron and nickel and is an average of 8800°F. It loses its heat to the mantle through convection, like boiling water in a pot.

The mantle is very different. It's temperature varies greatly at about 950°F to 1600°F. It makes up most of the mass of Earth. It is not liquid as most people think. It is solid. The asthenosphere contains melted magma plumes but overall the mantle behaves like silly putty. You pull it slowly and it flows, pull it fast and it snaps. It is mostly made of the mineral olivine. Like the outer core, the mantle transfers heat to the surface through convection. However, the process is very uneven and much slower than in the outer core. This leads to magma plumes near the core-mantle boundary making their way to the surface as hot spots. It also leads to cool solid parts as subduction brings the lithosphere into the mantle. The asthenosphere is now suspected to contain hydroxide ($OH^-$) bounded to minerals (like ringwoodite) and derived from water.

The lithosphere and crust operate as one unit. This is where the "plate" part in Plate Tectonics comes in. As a result of uneven mantle convection, the Earth's surface is like a cracked eggshell. It is divided into 12 large plates that move around the globe as magma from the mantle rises upward (divergent boundaries) causing the lithosphere to move laterally around the globe. In other areas the lithosphere is being subducted (convergent boundary) back into the earth, where it is reincorporated into the mantle, and the process starts over.

Gravity plays a major part in the plate tectonics cycle. As ocean crust ages, it becomes more dense and begins to sink into the mantle forming a subduction zone.

## Cross Section of the Earth

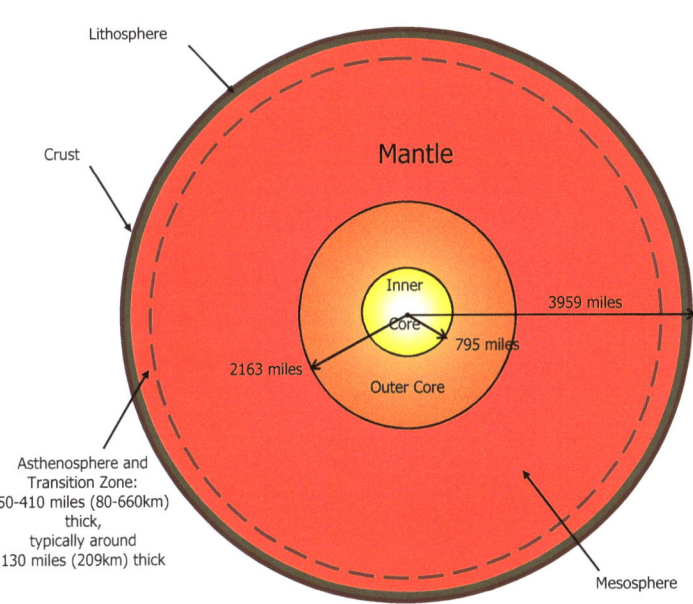

**PERCENT OF MASS (EARTH = 100%) AND AVERAGE DENSITY**

| Unit | Mass | Density | |
|---|---|---|---|
| | | (g/cm³) | (lbs/ft³) |
| Lithosphere + Crust | = 2.2% | = 2.8 | = 174.8 |
| Mantle | = 68.3% | = 3.3 - 6.0 | = 206.0 - 375.6 |
| Outer Core | = 27.5% | = 10.4 | = 649.3 |
| Inner Core | = 1.8% | = 13.3 | = 830.3 |
| Water | ≤ 0.2% | = 1.0 | = 62.4 |

# Igneous Parent Rocks and Metamorphic Equivalents

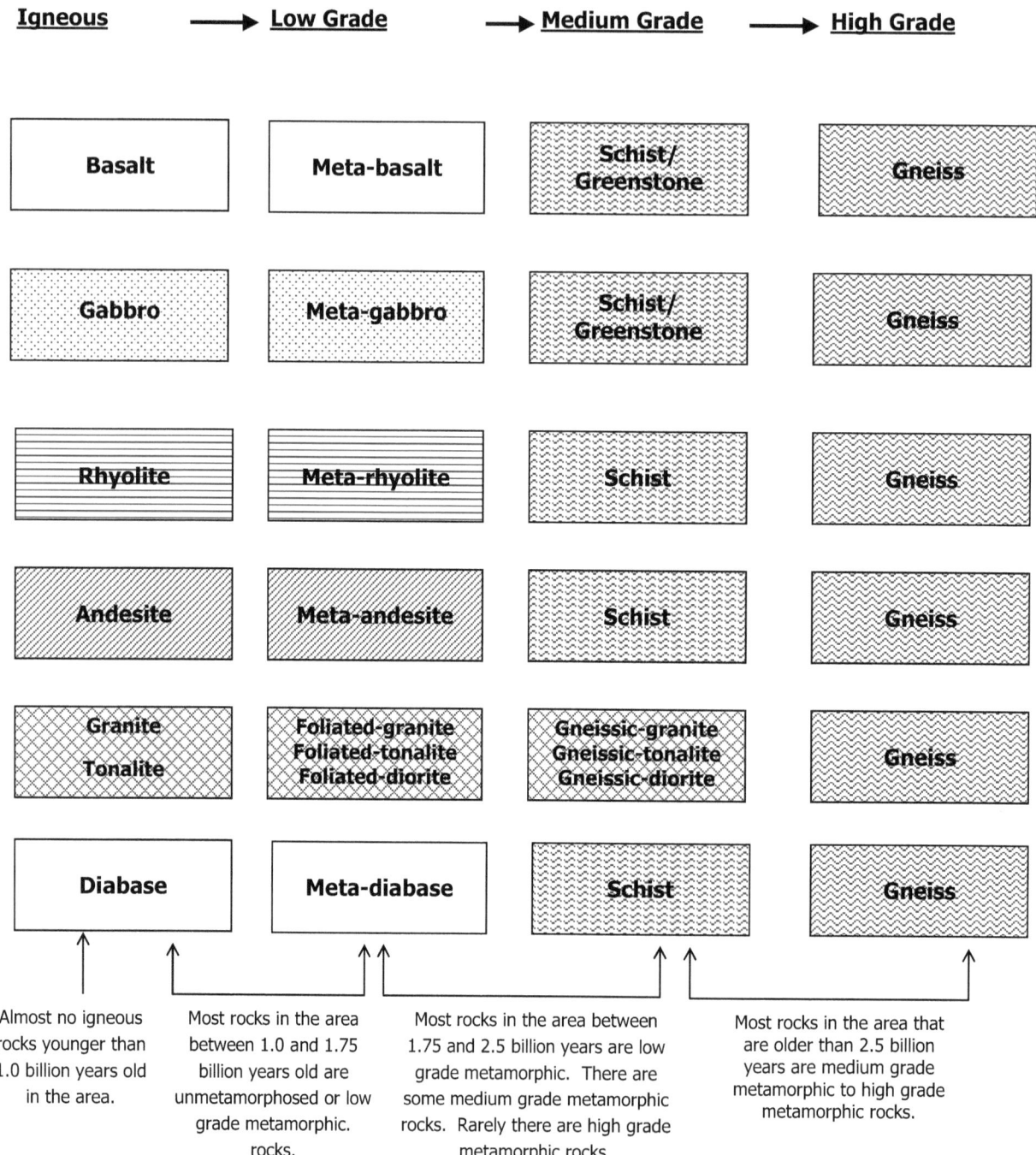

# Sedimentary Parent Rocks and Metamorphic Equivalents

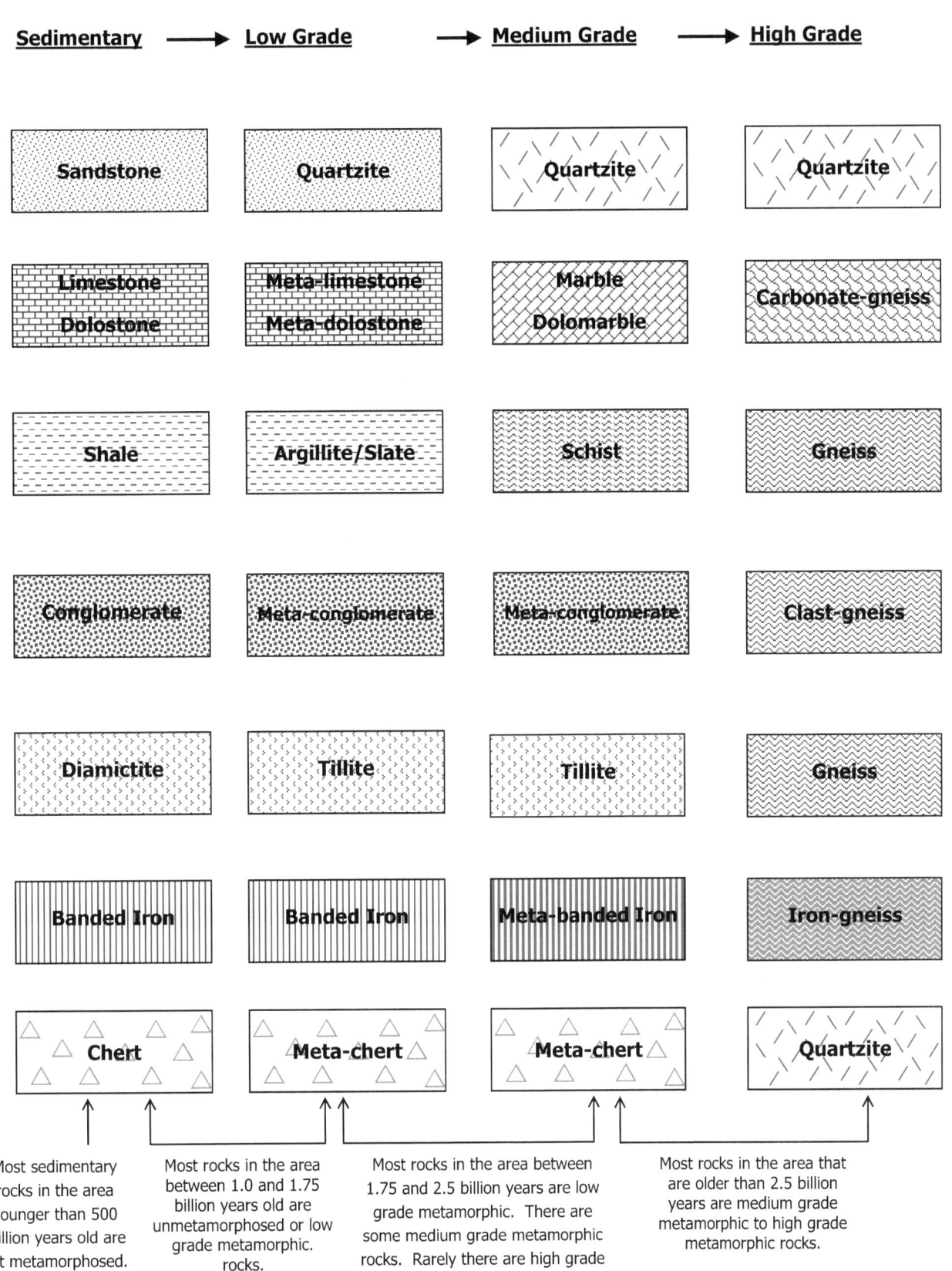

# Classification Chart of Coarse Grained Igneous Rocks

This chart is used for coarse grained (and usually plutonic) igneous rocks. Plutonic rocks tend to be coarser grained because they cool underground and thus crystalize more slowly than surface eruptions. This allows larger crystals to grow. It can be used in the field or the lab. Coarse grained in this case means the grains are easily visible to the naked eye. This corresponds roughly to 1/4 (0.25) millimeters (~0.01 inches) or the #60 U.S. and Tyler sieve sizes. This is the border between what geologists call fine and medium sized sand grains (based on the Wentworth Scale).

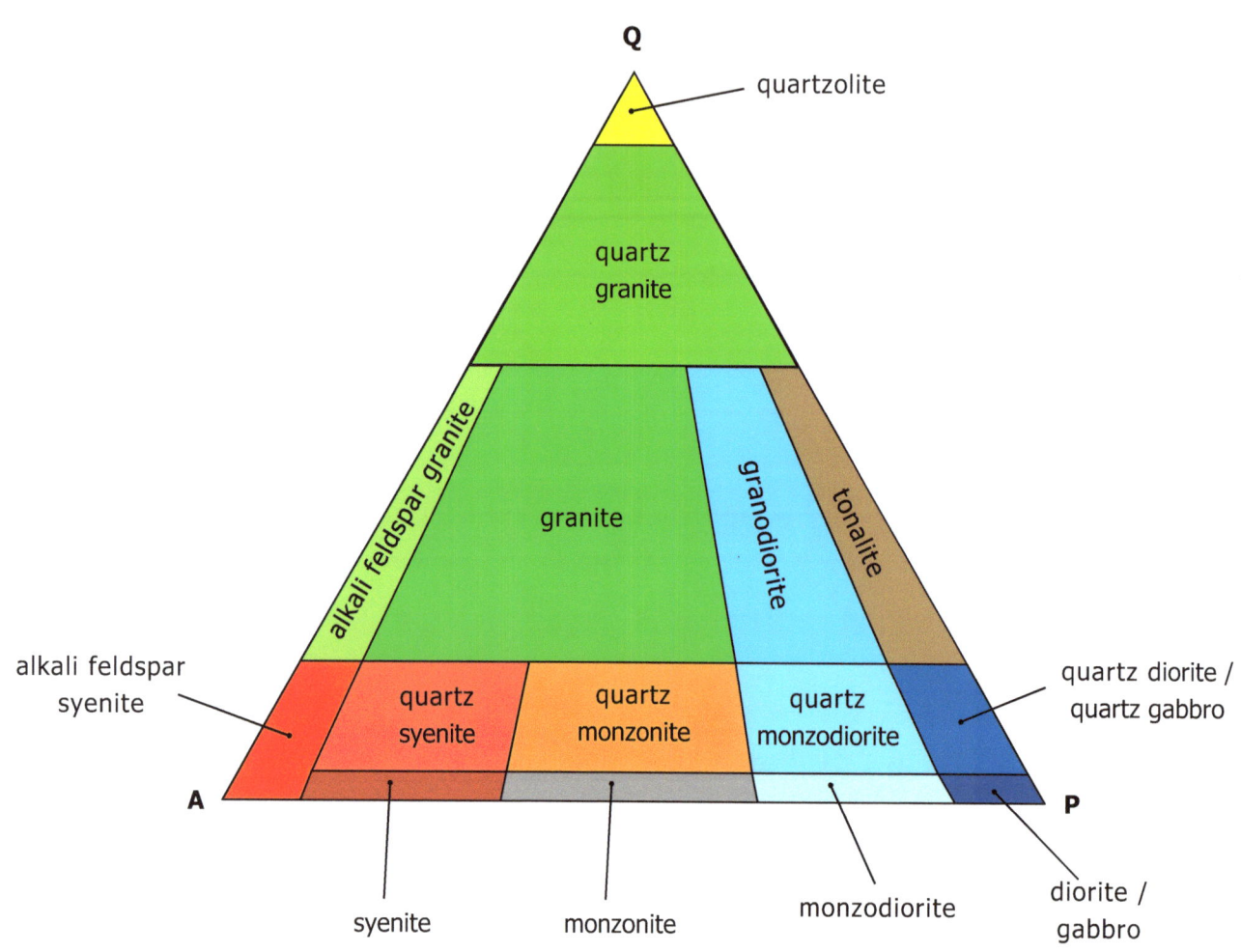

A = alkali feldspar
P = plagioclase
Q = quartz

*This chart ignores the dark colored minerals in coarse grained igneous rocks, such as hornblende and mica. It deals only in the ratios of quartz, plagioclase, and alkali (ak) feldspar. Dark mafic mineral percentages need to be removed before using this chart. If a plutonic rock contains >90% mafic minerals, this chart is not used.*

# Applicable Geologic Timescale

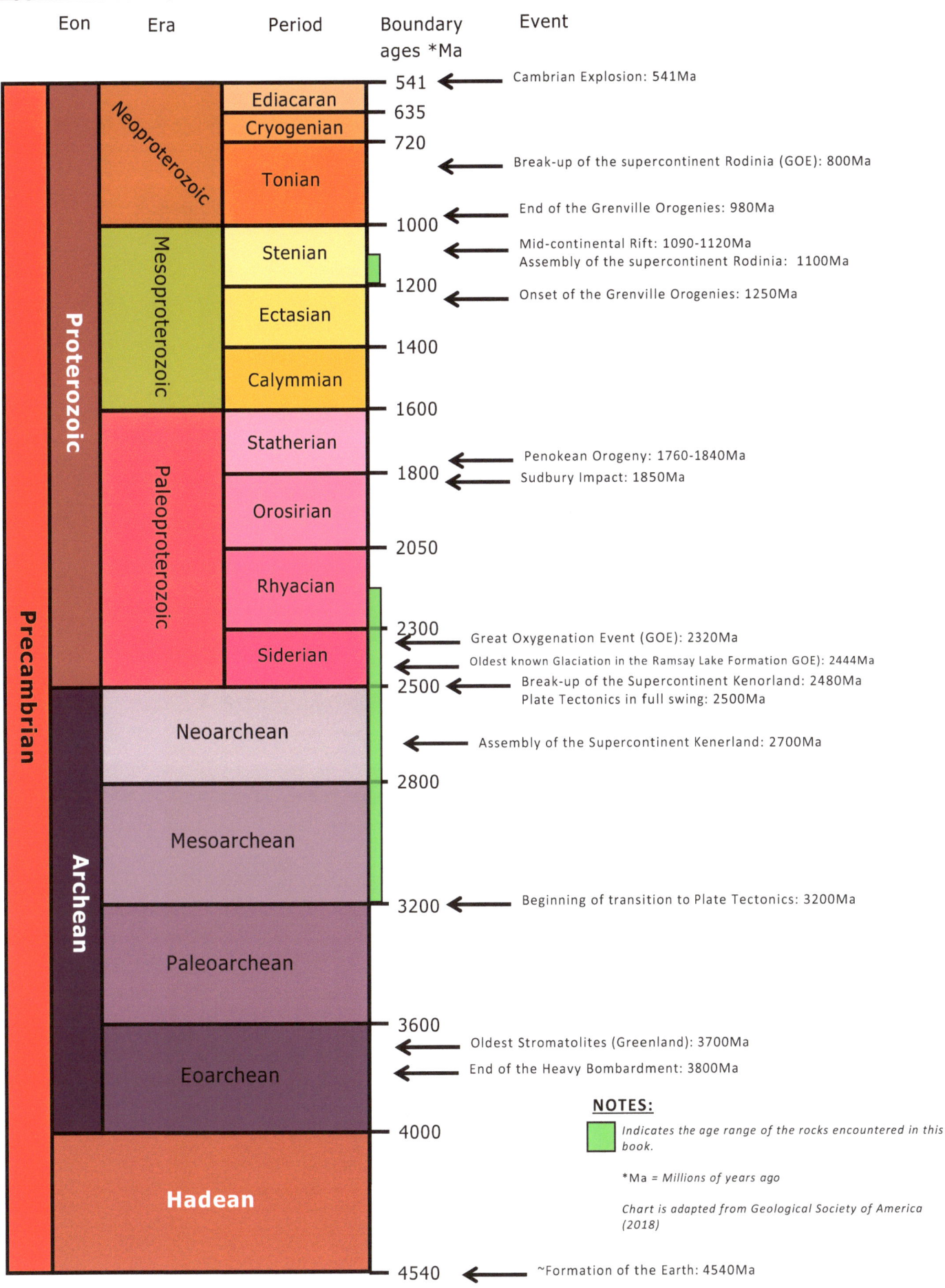

# Huronian Supergroup

The Huronian Supergroup is a thick and important assemblage of local rocks. This supergroup consists of 4 groups and 20 formations. The basal Elliot Lake group represents the formation of a continental rift. The Hough Lake Group represents the sediments that filled the rift zone as the ocean moved in. The Quirk Lake and Cobalt Groups represent the passive margin phase, similar to the modern North American Coastal Plain. The Nipissing represents intrusions that penetrated the Huronian as the passive margin became an area of active subduction. Not only does the Huronian demonstrate a modern style of Plate Tectonics, but it records the first free $O_2$ in the atmosphere as indicated by the appearance of the first "red beds" in the Firstbrook Member of the Gowganda Formation.

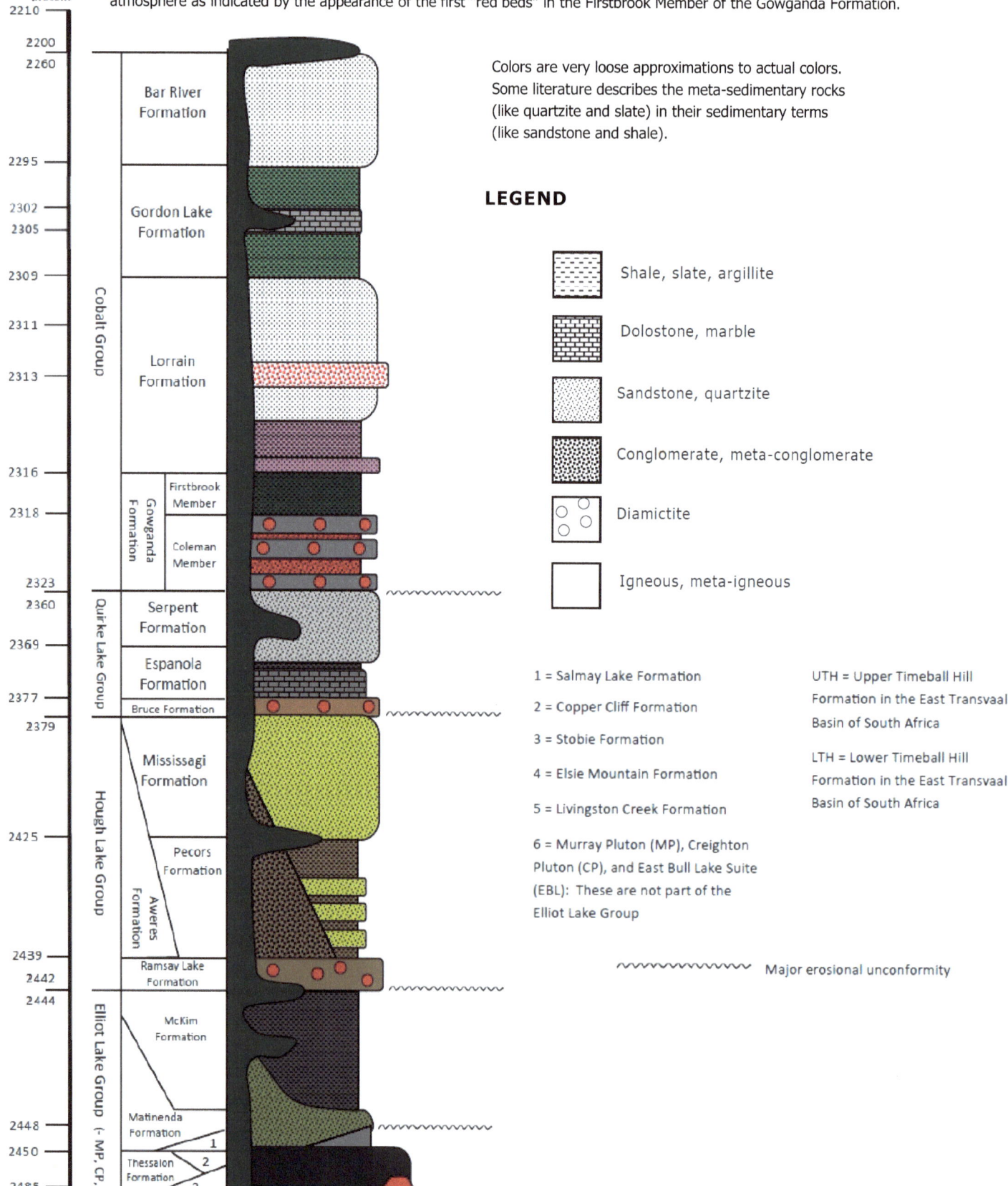

Colors are very loose approximations to actual colors. Some literature describes the meta-sedimentary rocks (like quartzite and slate) in their sedimentary terms (like sandstone and shale).

**LEGEND**

- Shale, slate, argillite
- Dolostone, marble
- Sandstone, quartzite
- Conglomerate, meta-conglomerate
- Diamictite
- Igneous, meta-igneous

1 = Salmay Lake Formation
2 = Copper Cliff Formation
3 = Stobie Formation
4 = Elsie Mountain Formation
5 = Livingston Creek Formation
6 = Murray Pluton (MP), Creighton Pluton (CP), and East Bull Lake Suite (EBL): These are not part of the Elliot Lake Group

UTH = Upper Timeball Hill Formation in the East Transvaal Basin of South Africa

LTH = Lower Timeball Hill Formation in the East Transvaal Basin of South Africa

~~~~~~~ Major erosional unconformity

# Major Types of Faults

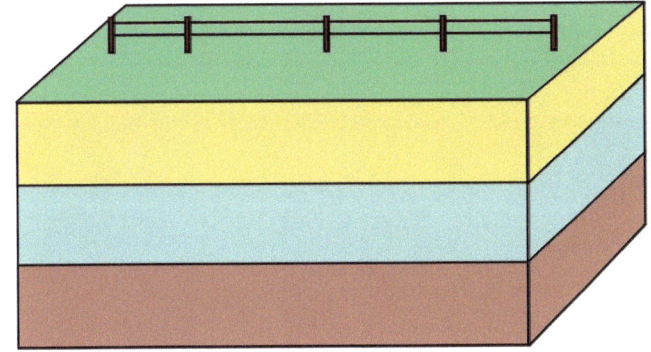

### Non faulted Block

This is a block diagram of a small section of the Earth. The vertical brown posts attached with dark lines represent a fence. The green is grass. The colored layers represent strata of different buried rock types.

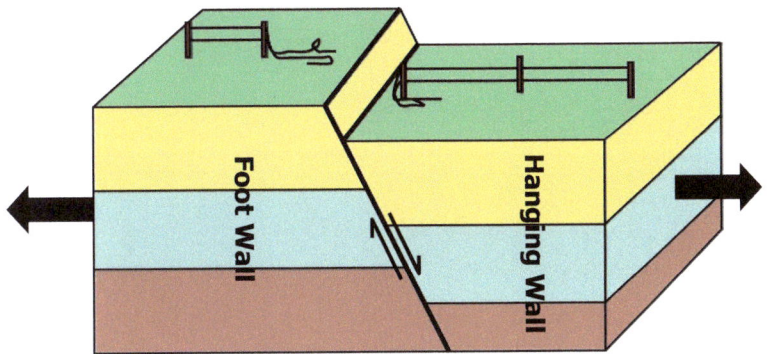

### Normal faulted Block

A normal fault forms when the Earth's crust extends, causing the hanging wall to drop down relative to the foot wall. Note the position of the broken fence. The small arrows indicate relative movement. The large arrows indicate the direction of extension.

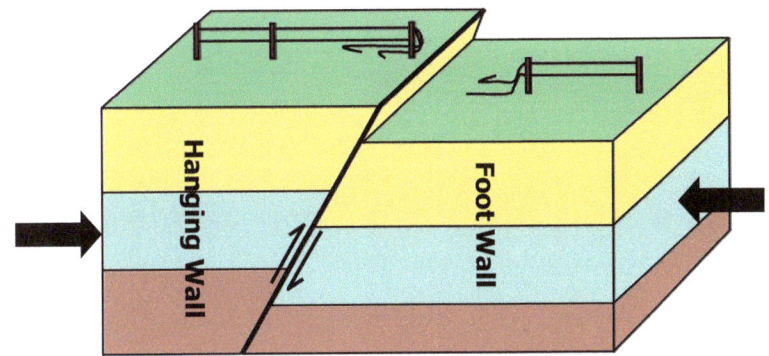

### Reverse faulted Block

A reverse fault forms when the Earth's crust compresses, causing the hanging wall to move up relative to the foot wall. Note the position of the broken fence. The small arrows indicate relative movement. The large arrows indicate the direction of compression.

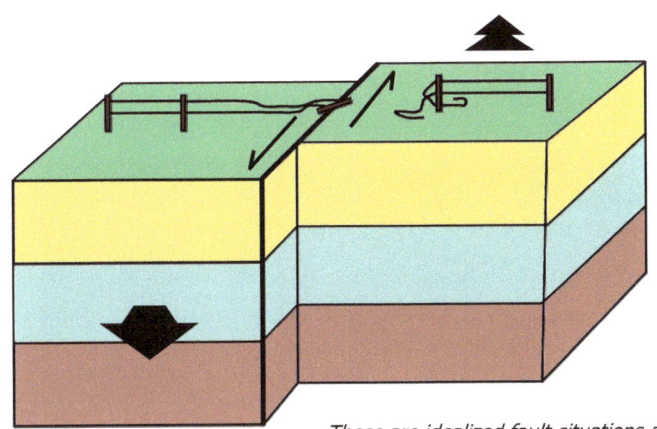

### Strike-slip faulted Block

A strike-slip fault forms when one block of the Earth slides laterally passed another due to transform movement. There is no hanging or foot wall. Note the position of the broken fence. The arrows indicate relative movement.

*These are idealized fault situations and are only the three main types. Most faults are far more complex and can be a combination of different fault types.*

# Large Scale Geologic Structures

Large scale geologic structures are mechanical alterations in the rock on a scale larger than a person. Some large scale structures, such as faults, can also be small scale. They can take up less than one square mile or thousands of square miles. They form after the sediment or rock has been deposited. They can be formed by tectonic movement, erosion, or intrusive magmas. The diagram below is a hypothetical (no scale) cross section, a slice through the Earth's crust. The colors below represent different types of rock.

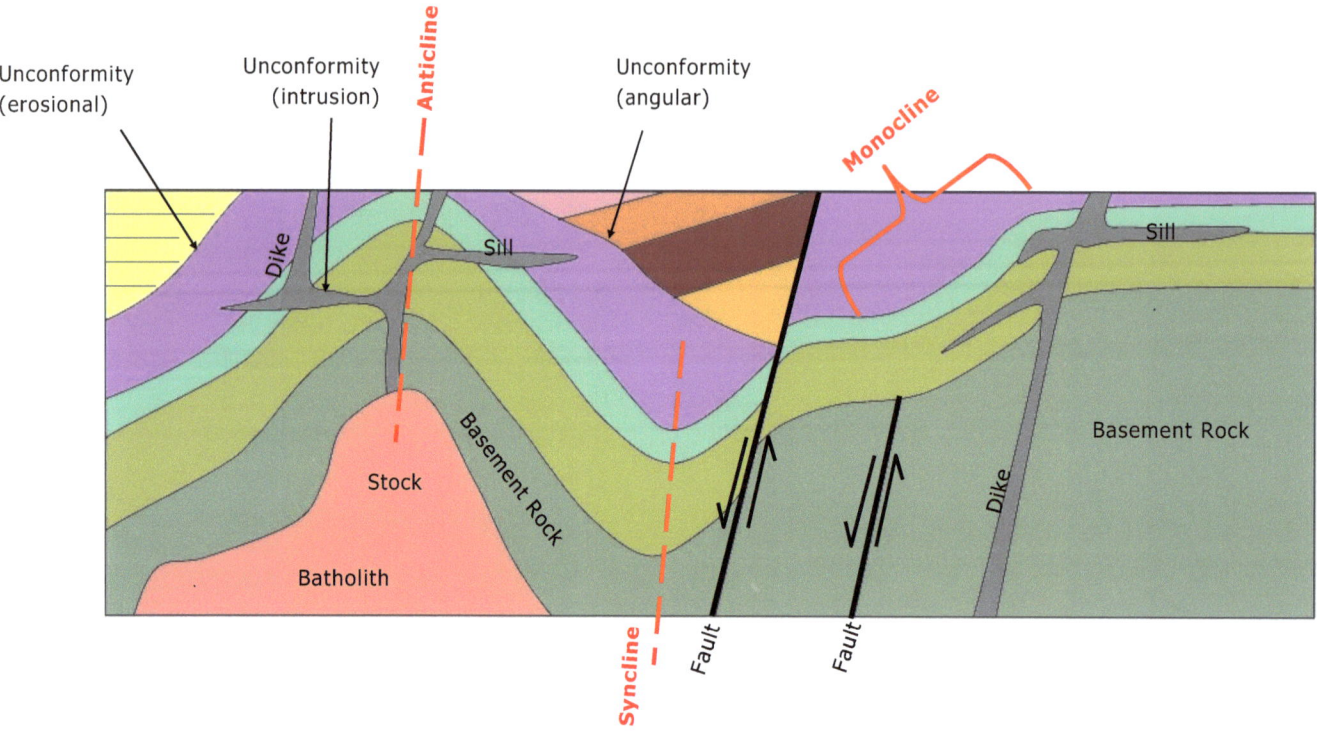

- **Anticline:** A folded set of rocks where an open ended ridge forms. This may or may not be reflected in the surface topography. Older layers are towards the center. An anticline that forms the shape of an upside down bowl is closed and referred to as a dome.
- **Basement Rock:** Usually refers to the oldest (Precambrian) rocks in an area. They can be anywhere from deep underground to at the surface.
- **Batholith:** A large igneous intrusion (usually several square miles) that was originally deposited by magma deep underground. Usually a coarse grained granite type rock.
- **Dike:** A vertical, narrow, igneous intrusion that has cooled from magma within existing rock.
- **Fault:** A relatively planar surface on which measurable movement has occurred.
- **Monocline:** A folded set of rocks, resembling a step, where only one end is steeply folded in the middle and the surrounding rock is relatively flat. It may be part of an anticline or syncline.
- **Sill:** A horizontal, narrow, igneous intrusion that has cooled from magma within existing rock.
- **Stock:** A small igneous intrusion (usually around one square mile or less) that was originally deposited by magma deep underground. Usually a coarse grained granite type rock. Stocks are often connected to batholiths.
- **Syncline:** A folded set of rocks where an open ended depression forms. This may or may not be reflected in the surface topography. Younger layers are towards the center. A syncline that forms the shape of a bowl is closed and referred to as a basin.
- **Unconformity:** An unconformity represents a missing block of time in the rock record, usually due to intrusions, erosion, or non deposition. There are many different types. An angular unconformity is the easiest type to spot in the field. Angular unconformities have one set of rocks at an angle to another set of rocks.

# How to Read a Geologic Map

Geologic maps are produced in order to represent geologic units on a base map (usually on a topographic map but can be any map) of a selected area. The purpose of a geologic map is to decipher and predict the relationships between geologic units in the selected map area. Most geologic maps represent the geologic units with colors (although some are black and white) and abbreviations of geologic units, which are explained in the legend. The map below is an excerpt from a map created by the author in 2013. The legend explaining the geologic units is not shown to conserve space. Maps presented in this book are accompanied with legends. Legends include not only the map symbols, but also geologic units. Features typically found on geologic maps of the area are shown below with arrows and circles. All geologic maps must contain three things. 1-north arrow. 2-scale. 3-map legend. The United States Geological Survey has set standards for geologic symbols such as contacts, strike and dip, faults, anticlines, and other structures not depicted below. Colors are also standardized. However, if you have a lot of geologic units, especially when most of them are within the same span of geologic time, the colors would be so close to one another that it is acceptable to deviate from the standard. For example, if you had 13 geologic units, all of which are Paleoproterozoic, it wouldn't make sense to have 13 shades of purplish red on your map.

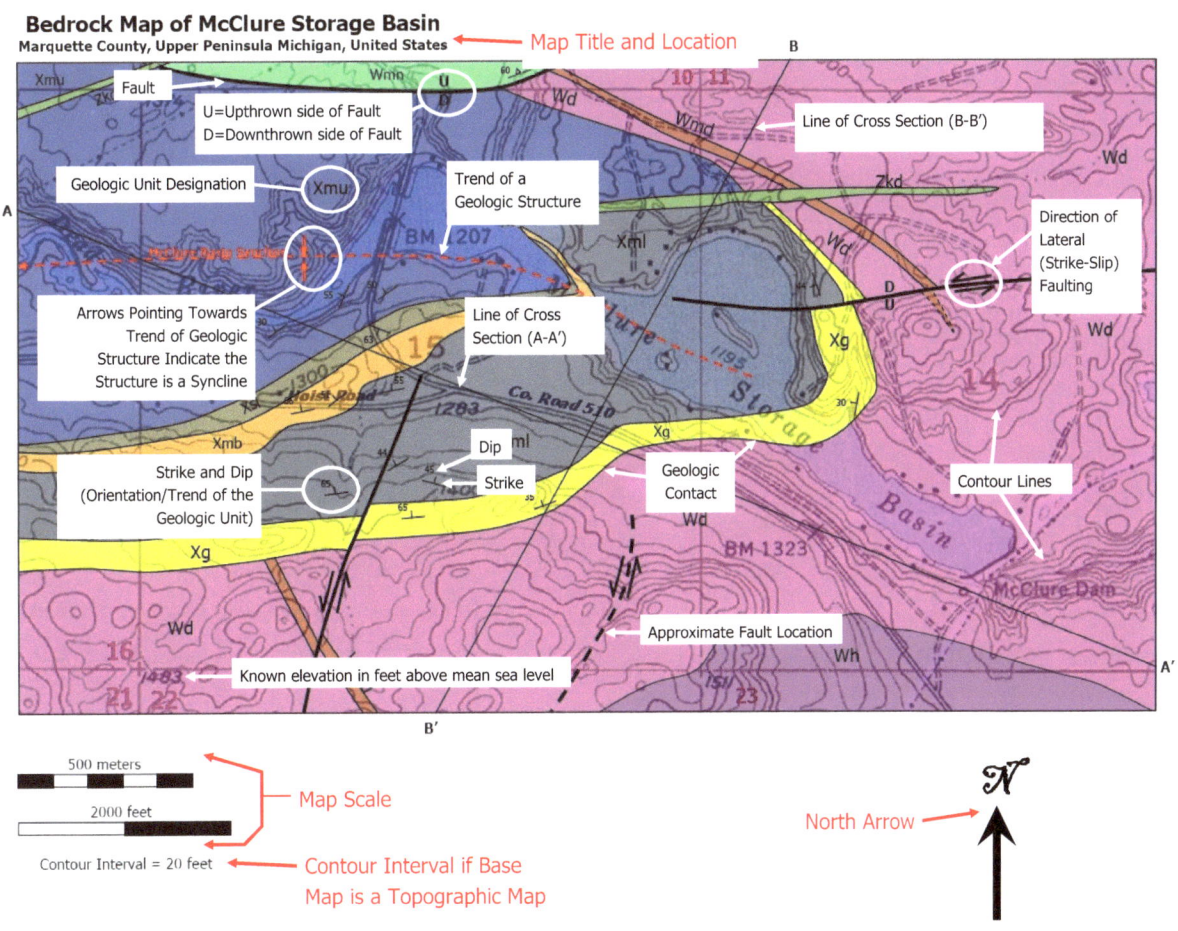

# Bowen's Reaction Series

This is an important chart that illustrates how minerals form as they cool from magma. Magma is not made of minerals. It is a chemical liquid from which minerals form. Minerals by definition are solids. Bowen's reaction series chart assumes a magma body with all the available elements to form minerals on the chart as the magma cools. If an element weren't present, say, iron, you would not get olivine. The chart is an idealized reaction series. It does not take into account things like two chemically different magmas merging, rapid verses slow cooling, fragmentation of the magma, chemical enrichment of the magma due to hydrothermal or gas injection, or incorporation of abundant and chemically distinct country rock into the magma as it rises and cools.

At the dawn of the 20th century, Norman L. Bowen and others, began experimenting with rocks in order to see if certain minerals would crystalize from a magma first. Several things were discovered out of the experiments. First, the magma (or melt) would try to stay in equilibrium with the forming crystals (or minerals). This would result in different compositions between the melt and minerals, and would through off equilibrium. So the new crystals would re-react with the melt to form new minerals. Second, the newly crystalized minerals would form in a specific order. Third, if the magma had enough silica and was homogenous two main series would form. The continuous and discontinuous series, which would both merge to a simpler chemical melt composition of the magma as the iron, magnesium, calcium, sodium, etc. was used up to form the residual series.

The continuous series deals with the crystallization of the feldspars. First plagioclase minerals will form. As the magma cools, this will through off the equilibrium. So some of the calcium in the plagioclase will re-react with the magma and become potassium/sodium enriched until finally orthoclase crystalizes out. The discontinuous series is odd...but makes sense. Say the magma produces olivine at a high temperature. As it cools further the olivine reacts with the melt but doesn't exchange ions like in the continuous series. Instead it will "change" into pyroxene. The chemical reaction would look something like this (from a generic olivine to pyroxene): $Mg_2SiO_4 + SiO_2 \rightarrow 2MgSiO_3$. So, instead of say calcium being replaced with potassium, the added silica causes the existing mineral to reorganize itself but made of the same chemicals. Basically all that is occurring is there is an internal crystal lattice (or structural) adjustment to achieve crystalline stability at lower temperatures until orthoclase is formed. Whether continuous or discontinuous, once orthoclase forms, only the residuals remain. Muscovite will form until everything but silica ($SiO_2$) is left to form quartz.

How does this help us identify minerals? For starters, olivine and anorthite can form together as they are both mafic minerals. But pyroxene and quartz cannot form together since one is mafic and the other is felsic. Say you have a fine to medium grained igneous rock and you see about 50% very dark but unidentifiable mineral, and 50% nearly white and heavily striated mineral. You can't identify the dark mineral but it can be assumed to be pyroxene. Why? Because the other mineral is white and heavily striated, so it has to be anorthite. Both are mafic so both can form together! Be careful. Just because anorthite and quartz can't form at the same time, that doesn't mean they can't be in the same rock as quartz can still form at low magma temperatures if the magma is saturated in silica.

As ingenious as Bowen's reaction series is, it cannot account for everything. In Bowen's day, we didn't know about plate tectonics. The rolls of saturated verses unsaturated magma, incorporation of country rock, differentiation of magma due to cumulate was only beginning to be studied. During his time people were still arguing if granite was igneous or metamorphic. Bowen's reaction series is very useful but it deals with idealized conditions. This is why field mapping in conjunction with laboratory petrographic analysis is so important. The two cannot exist in isolation.

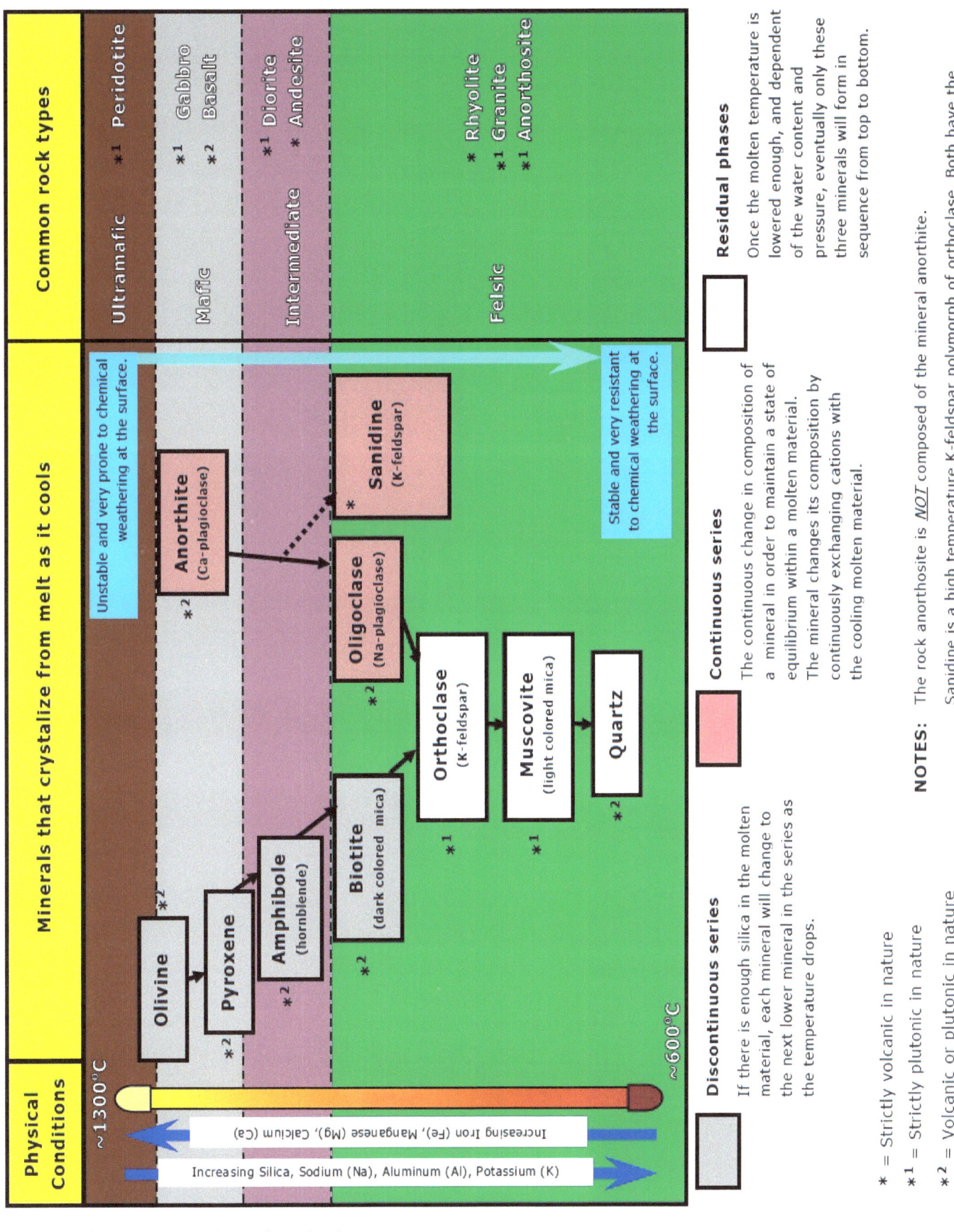

*Adapted from Bowen (1922) and others

# Outcrop Stops: Location Maps

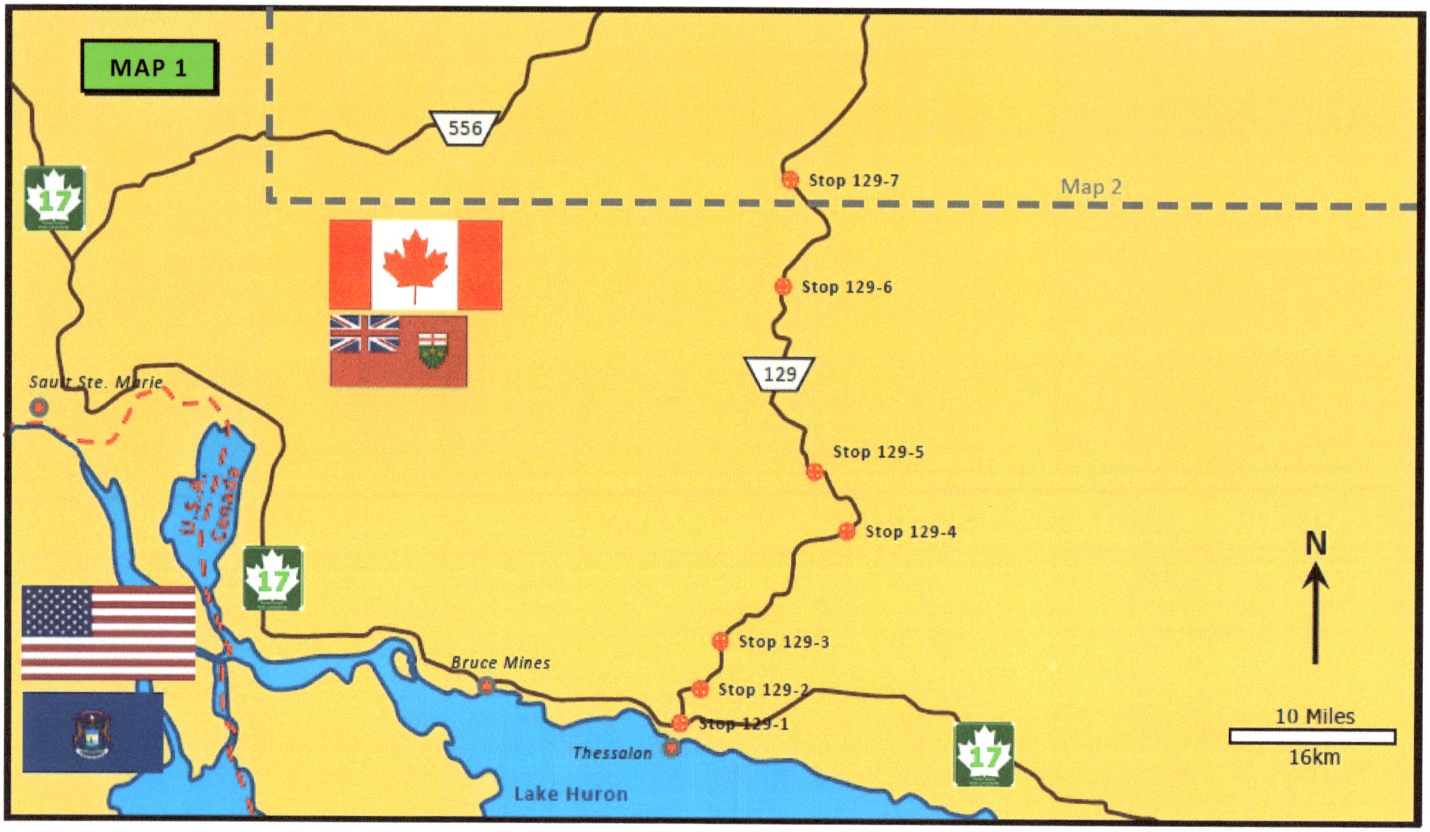

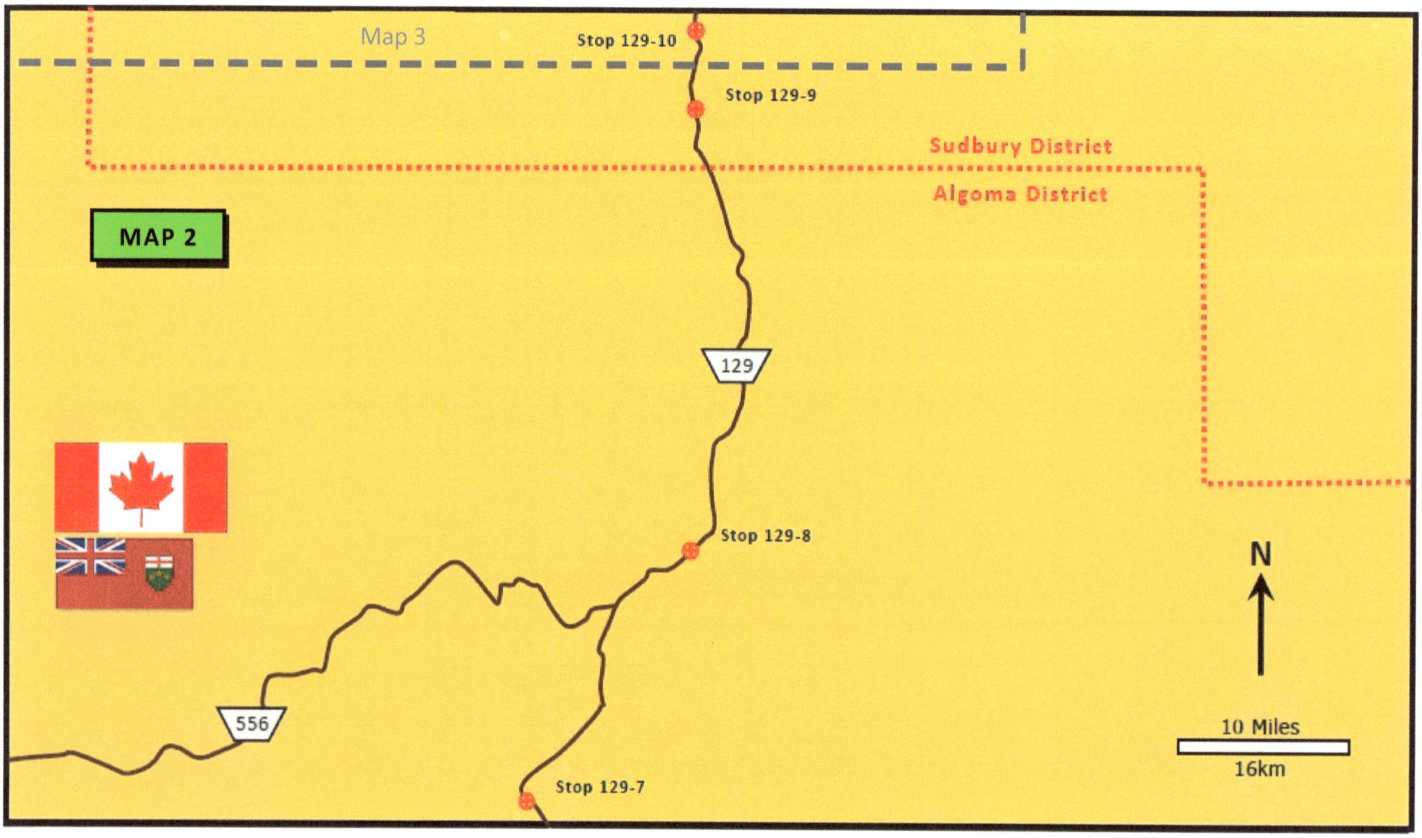

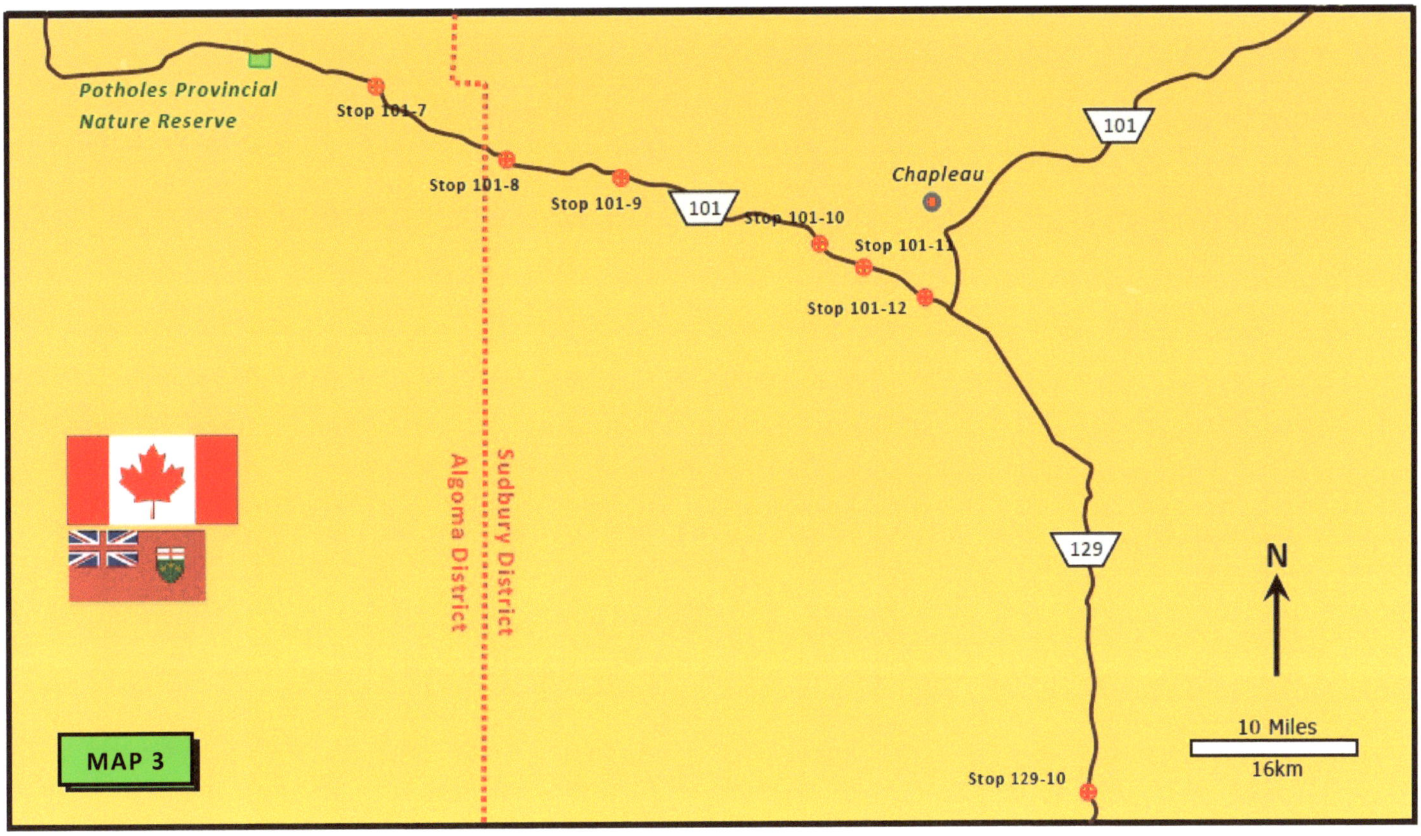

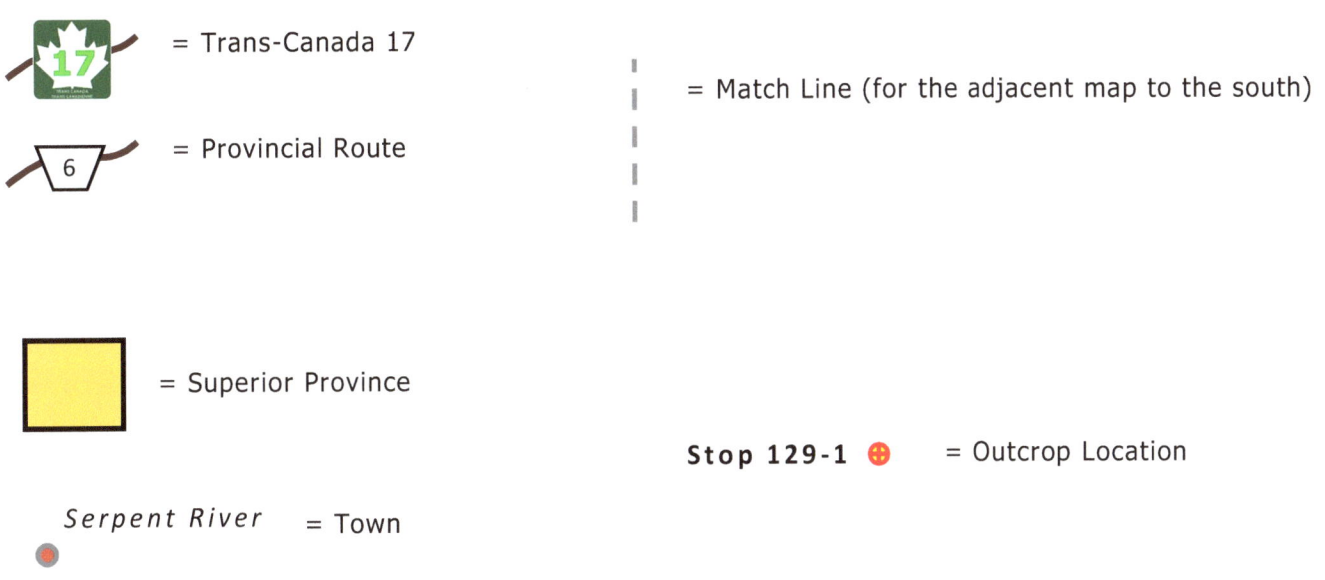

**LEGEND:**

- = Trans-Canada 17
- = Provincial Route
- = Match Line (for the adjacent map to the south)
- = Superior Province
- Stop 129-1 = Outcrop Location
- *Serpent River* = Town

NOTE: These maps are approximations to give you a general idea of how the stops relate to one another. It is not an exact representation of the ground surface. It **does not** depict all roads, towns, topography, all small lakes, or all rivers.

# Map to Current Roadside Books of Ontario

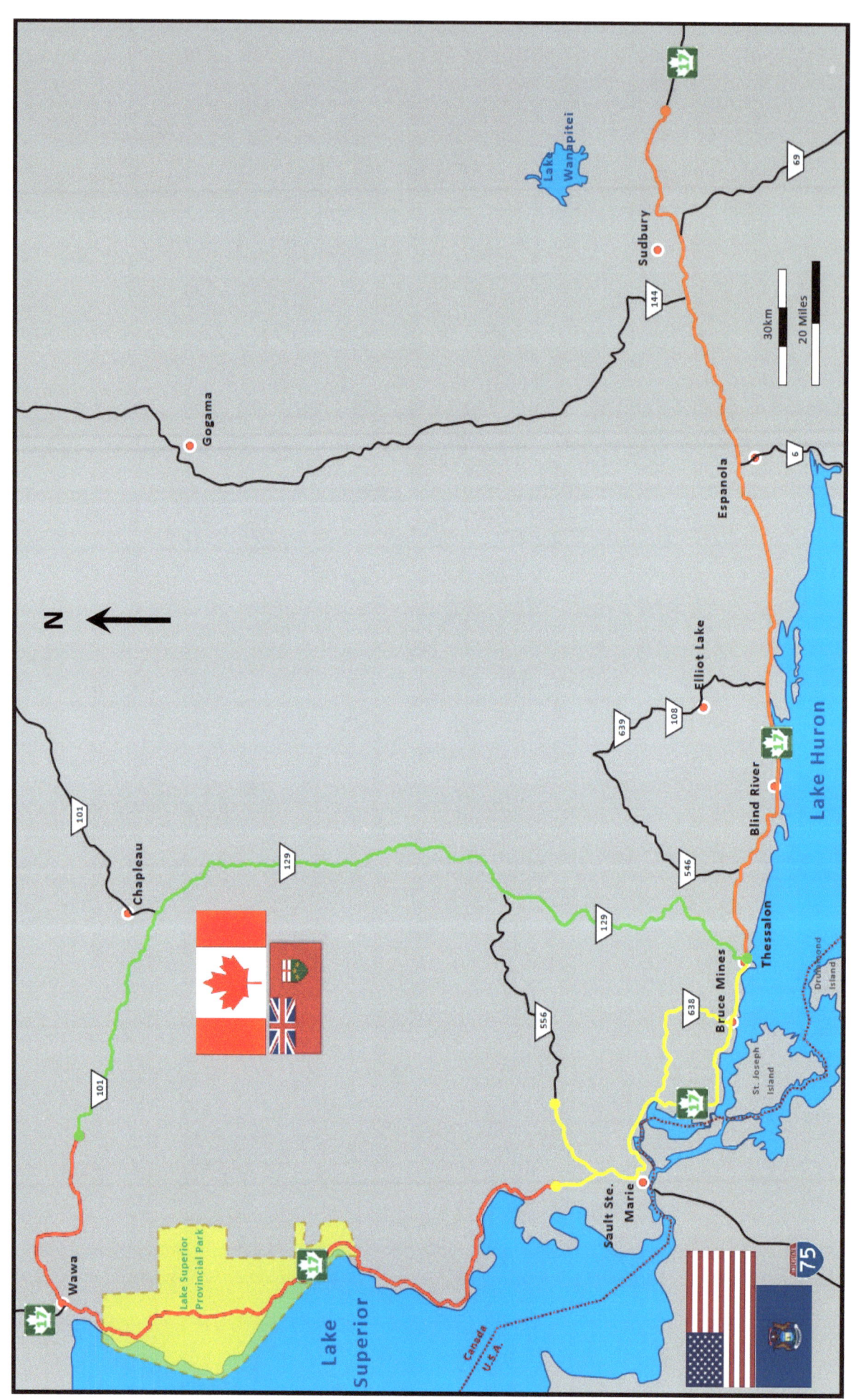

**LEGEND:**

**Roadside Geology of the Algoma District, Ontario: From the Goulias River, to Wawa, to Potholes Nature Reserve**

(Roadside Geology of the Midwest, Volume 4)

**Roadside Bedrock Geology Along Trans-Canada 17: From Thessalon to Sudbury, Ontario**

(Roadside Geology of the Midwest, Volume 5)

**Roadside Geology of the Trans-Canada 17 and Route 638 Loop in Ontario Canada: From Goulais Bay to Thessalon**

(Midwest Roadside Geology, Volume 1)

**Roadside Bedrock Geology Along Route 129 and 101: From Thessalon to Potholes Provincial Nature Reserve, Ontario**

(Roadside Geology of the Midwest, Volume 6)

Roads not included in any roadside book

17 = Trans-Canada Route

INTERSTATE 75 = U.S. Interstate Route

6 = Provincial Route

= International Border

**Sault Ste. Marie** = Town / City

**OUTCROP NAME:** Route 129 and Trans-Canada 17 Outcrop  **OUTCROP DESIGNATION:** 129-1

**OUTCROP LOCATION:** GPS: 46.26678 −83.54835  **ELEVATION:** 610 feet (185.9 meters) above Mean Sea Level

**FORMAL GEOLOGIC NAME:** Thessalon Formation

**MAIN ROCK TYPE(S):** Meta-gabbro with epidote phenocrysts

**DESCRIPTION:** The outcrop is located 547 (166.7 meters) north of Trans-Canada 17 (TC-17) along Route 129.

Here the rock is a dark gray to greenish gray, slightly metamorphosed gabbro. Crystals are poorly developed. The rock is medium equigranular (meaning roughly equal crystal sizes). Compared to other gabbros in the area, this one is much finer grained. Some geologists would call this a basalt. Even though the crystals are smaller than most, they can be seen with the unaided eye and are still identifiable with a 10x hand lens. The crystals in a basalt are never visible with the naked eye.

It contains numerous rust colored small veins in fractures. There are three main fracture sets: ~N85E20SE, N25E85SE, and N83E25NW. There is a prominent planar white quartz vein that trends N81W74NE.

Part of the rock contains irregular shaped light yellowish green epidote phenocrysts about 0.5 inches (1.3cm) to about 1.3 inches (3.3cm). These phenocrysts likely formed from later hydrothermal alteration of the pyroxene in the gabbro, not from metamorphism.

The top of the outcrop is rounded due to the Quaternary ice sheets passing over. Small glacial striations can be seen on its top.

**FIGURE: Geologic Map**

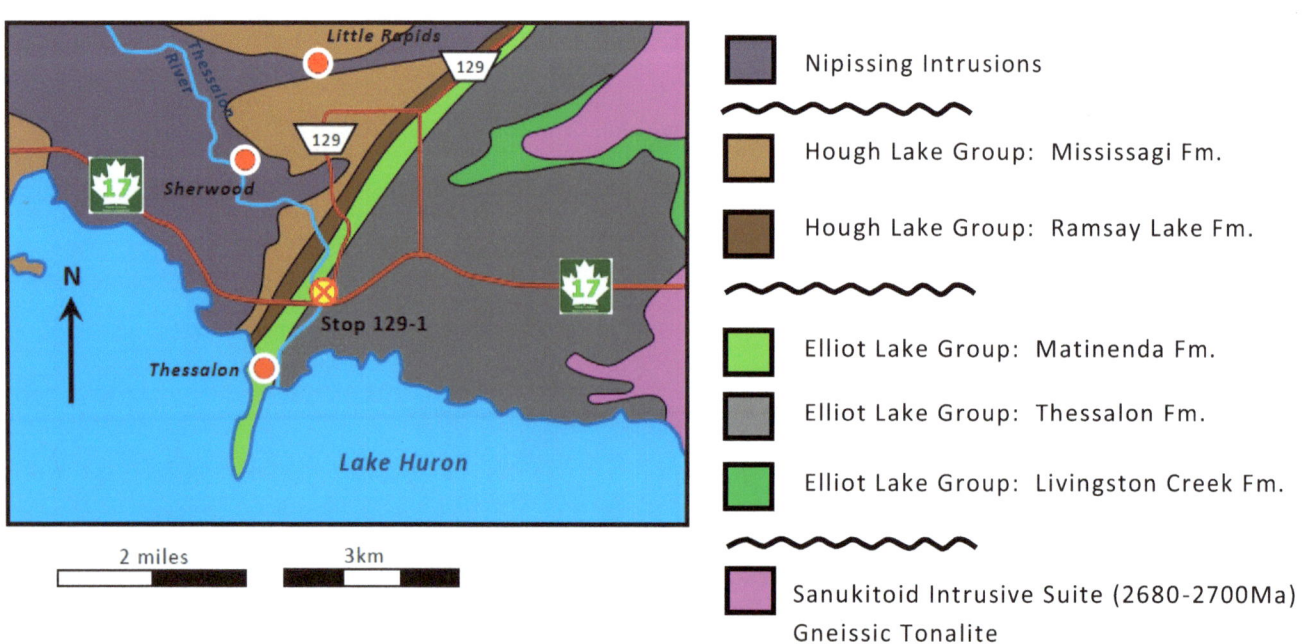

**PHOTOS:**

Outcrop, looking south. 5'11" (1.80 meter) person for scale.

Close-up of meta-gabbro, looking east. $2 piece scale.

Close-up of white quartz vein that trends N81W74NE, looking east. $2 piece scale.

Close-up of epidote phenocrysts (yellow arrows), looking east. $2 piece scale.

**OUTCROP NAME:** Route 129 and Station Road Northeast Outcrop  **OUTCROP DESIGNATION:** 129-2

**OUTCROP LOCATION:** GPS: 46.29878 –83.52058

**ELEVATION:** 655 feet (185.9 meters) above Mean Sea Level

**FORMAL GEOLOGIC NAME:** Ramsay Lake Formation

Matinenda Formation

**MAIN ROCK TYPE(S):** Proto-quartzite with red jasper clasts

**DESCRIPTION:** The outcrop is located 1693 feet (0.321 miles) or 516 meters northeast of the intersection with Station Road along the northwest side of Route 129.

The rock is almost entirely a gray medium to coarse crystalline proto-quartzite. There are red jasper inclusions spread randomly throughout, but are more common on the northeast end of the outcrop.

This outcrop is really close to the Ramsay Lake-Matinenda contact. As a matter of fact, on the southwest end of the outcrop, the contact is visible, a may be a fault contact. The contact dips 67° to the west. When looking at the outcrop, the left side of the contact belongs to the Matinenda and is dark and argillaceous. While to the right of the contact the rock is a light colored mottled purple proto-quartzite of the Ramsay. Rock beds on both side of the fault dip 20° to 30° to the northwest.

**FIGURE: Geologic Map**

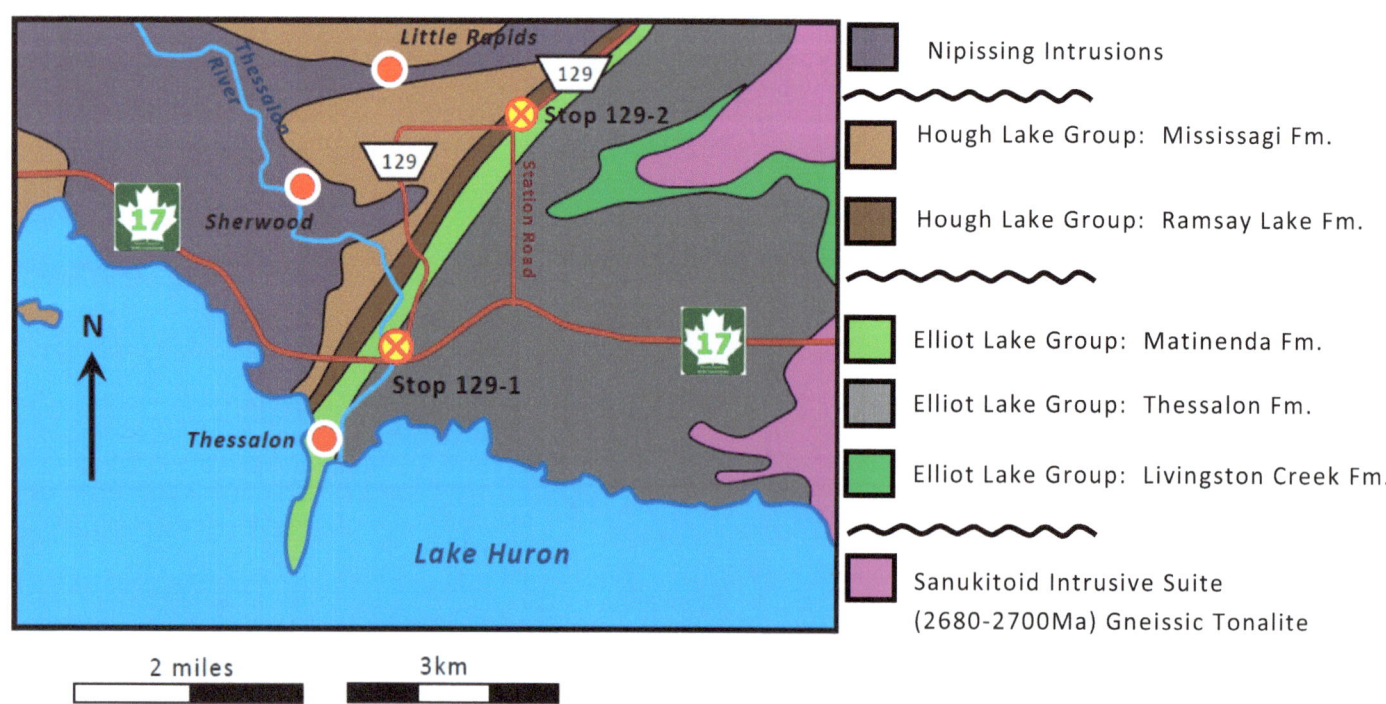

**PHOTOS:**

Outcrop, looking west-southwest. 5'10" (1.78 meter) person for scale.

Close-up of proto-quartzite of the Ramsay Lake, looking northwest. Arrows pointing to red jasper clasts. $2 piece scale.

Close-up of proto-quartzite of the Ramsay Lake, looking northwest, at the northeast end of the outcrop. $2 piece scale.

**OUTCROP NAME:** Route 129 North of Yates Lane Outcrop

**OUTCROP DESIGNATION:** 129-3

**OUTCROP LOCATION:** GPS: 46.33636 –83.49794

ELEVATION: 821 feet (250.2 meters) above Mean Sea Level

**FORMAL GEOLOGIC NAME:** Nipissing Intrusions

**MAIN ROCK TYPE(S):** Monzogabbro

**DESCRIPTION:** The outcrop is located along both sides of Route 129 about 860 feet (262 meters) north of the intersection with Yates Lane (dirt road).

Other than the Murray Fault, several clicks to the south, there are no known faults in the vicinity of this outcrop. We only had time to visit one of the two exposures, so we opted for the one on the east side of Route 129.

No contacts with the surrounding meta-sedimentary rocks were observed but this outcrop is igneous. It is pretty typical of the gabbro type rocks that make up the diverse package of the 2.2 billion year old Nipissing Intrusions. Overall the rock is a monzogabbro. It contains about 15% to 25% white plagioclase that is mostly micro crystalline. The remaining parts of the rock are medium to coarse weakly developed very dark green pyroxene with some coarse black hornblende crystals. There is no primary quartz in the rock. There are some white secondary, very thin quartz veins. There are places within the rock, where hornblende is more abundant at the expense of the pyroxene.

**FIGURE: Geologic Map**

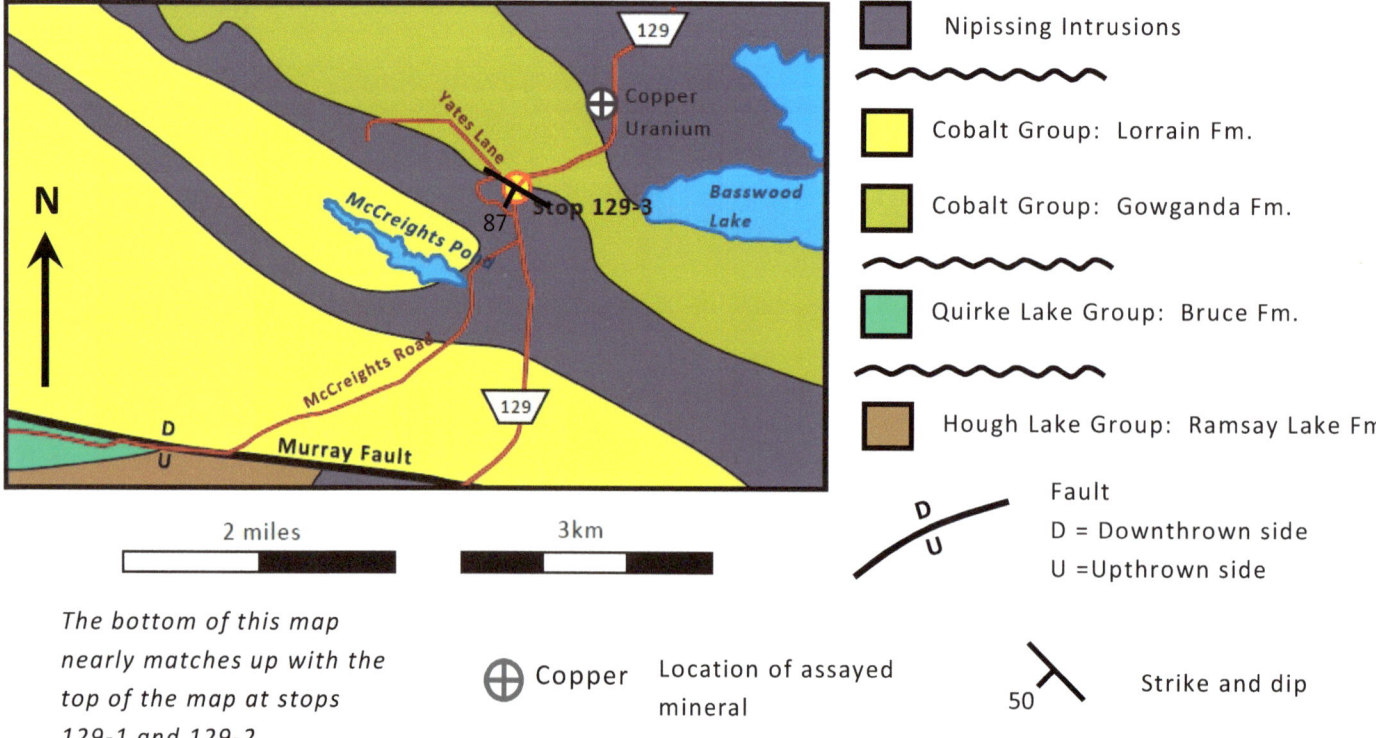

*The bottom of this map nearly matches up with the top of the map at stops 129-1 and 129-2.*

**PHOTOS:**

Outcrop on both sides of Route 129 (only the east side was studied), looking north-northeast. No scale.

Studied outcrop, looking east. 5'10" (1.78 meter) person for scale.

Close-up of monzogabbro, looking east. The white plagioclase along with the black pyroxene and hornblende are visible. $2 piece scale.

**OUTCROP NAME:** SE Side Route 129 East of Appleby Lake Outcrop  **OUTCROP DESIGNATION:** 129-4

**OUTCROP LOCATION:** GPS: 46.42765 –83.34312

**FORMAL GEOLOGIC NAME:** Gowganda Formation

ELEVATION: 813 feet (247.8 meters) above Mean Sea Level

Coleman Member

**MAIN ROCK TYPE(S):** Diamictite (tillite), Proto-quartzite, granitic clasts

**DESCRIPTION:** The outcrop is located on the south side of Route 129, just opposite of Appleby Lake.

The rock here is the diamictite (or tillite) of the Gowganda formation. The basal Coleman Member makes up the entire outcrop. The Coleman is the part of the Gowganda that contains the Paleoproterozoic glacial deposits.

The host rock of the tillite is a dark gray, unsorted, massive, argillaceous slate. It contains about 1% to 5% red granitic clasts that very in size from rounded granules to subrounded cobbles. They were derived from the Archean rocks to the north and northeast.

At this outcrop is a red proto-quartzite whose parent rock was a medium grained, lithic arkose. It isn't clear if this proto-quartzite was a sandy lens in the till of the Gowganda or an injected clastic dike. The former is more likely. The red proto-quartzite trends N65E84SE and is roughly 4 to 5 feet (1.2 to 1.5m) thick. Whether or not the lens is representative of bedding is something that is difficult to determine since the tillite lacks internal structure.

There are nearly horizontal slickensides that trend about N10E present on some surfaces. These likely formed when the adjacent Nipissing rocks were intruded. Although the Gowganda is extremely close to the Nipissing here, it shows no hydrothermal activity or metamorphism due to the Nipissing.

**FIGURE: Geologic Map**

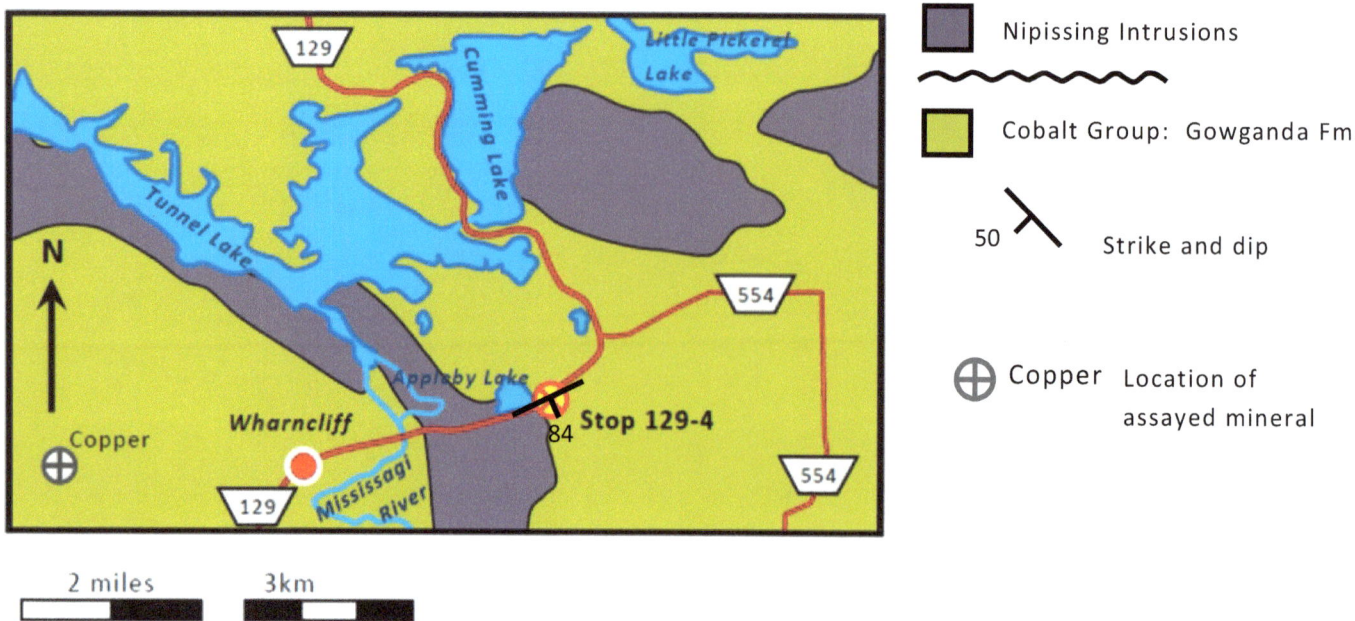

**PHOTOS:**

Outcrop, looking southwest. 5'10" (1.78 meter) person for scale.

Outcrop, looking northeast along strike of the red proto-quartzite, which trends N63E84SE. Jeep Patriot for scale.

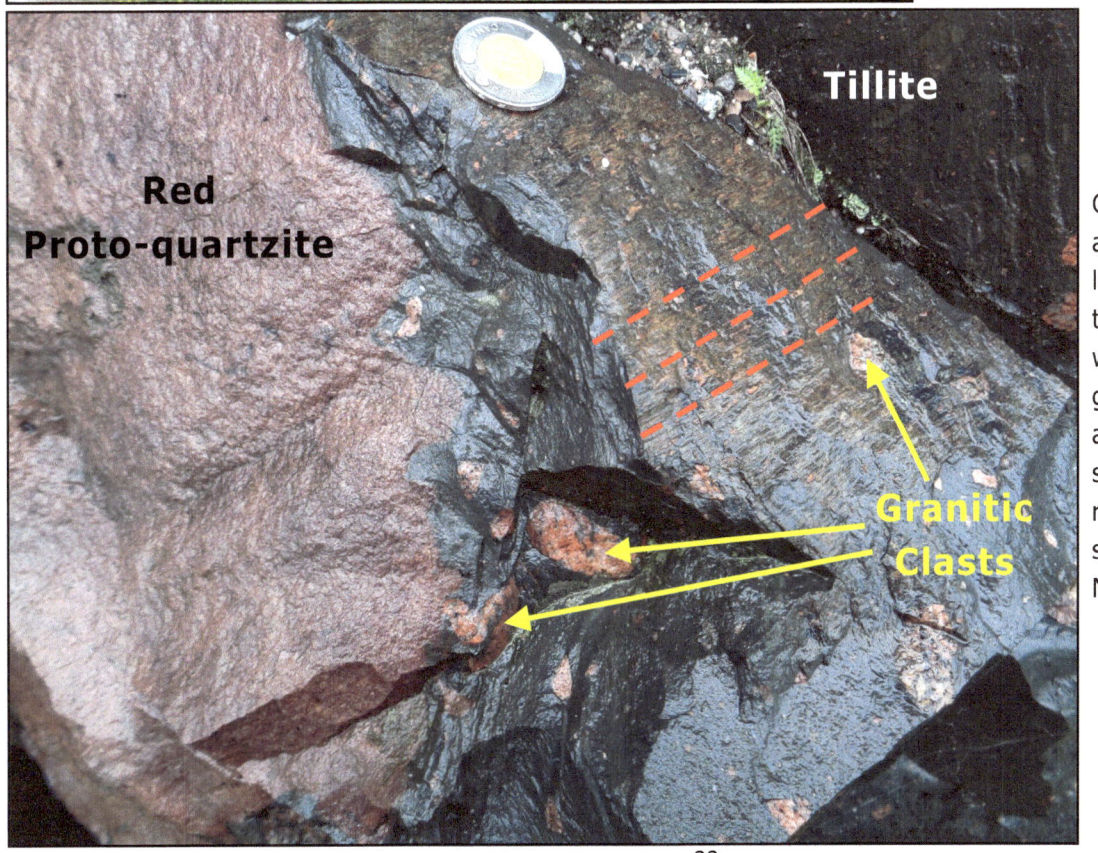

Close-up of the tillite and proto-quartzite, looking southeast. The tillite is the dark rock which contains red granitic clasts (yellow arrows are pointing to several of many). The red dashed lines are slickensides that trend N10E. $2 piece scale.

**OUTCROP NAME:** North Side of Tunnel Lake Outcrop

**OUTCROP LOCATION:** GPS: 46.47762 –83.38628

**FORMAL GEOLOGIC NAME:** Gowganda Formation

Coleman Member

**MAIN ROCK TYPE(S):** Proto-quartzite, granitic clasts

**OUTCROP DESIGNATION:** 129-5

ELEVATION: 1011 feet (308.2 meters) above Mean Sea Level

**DESCRIPTION:**

The outcrop is located on the north side of Route 129, between Tunnel and Cumming Lake.

There are outcrops on both sides of Route-129, but we studied the one on the north side of the road. The red rock here stands out in stark contrast to the surrounding nearly black outcrops of the area. Like at stop 129-4, the red rock is a proto-quartzite, except it is the entire outcrop. It is a medium to coarse granular to crystalline proto-quartzite. The parent rock is an arkose. There are almost no silt or clay sized particles. Overall the rock is massive, but faint laminations are occasionally present indicating a trend of N42W9SW.

The rock does contain isolated granitic clasts, just like the tillite of the Gowganda. This rock was likely deposited at the base of a Proterozoic glacier, or as outwash. The granitic clasts tend to be larger and rarer than in the tillite. This leads me to think that proto-quartzite is outwash.

Stratigraphically, this outcrop is likely near the top of the Coleman Member. It makes sense that sand would be deposited at the top of the sequence as the glaciers retreated.

The outcrop shows an excellent textbook joint face. Joints form when a rock breaks off from another at a single point. Joints commonly form as freeze-thaw cycles cause the rock to break.

**FIGURES: Basic Joint Face Diagram**

This is a basic diagram of a joint face. This diagram only depicts the basic structure. (There can be other parts not depicted). Joints form from catastrophic breaking of a rock from a single point (the origin) in a fraction of a second. This can occur naturally or artificially (like breaking it with a hammer). The overall structure formed is called a Plumose Structure and is made of several parts. As the shockwaves propagate from the origin (blue arrow), ribs (gray arcs) will form ring patterns that represent a brief millisecond slowing in the shockwave as it moved through the rock. Ribs may be very faint or not present if the density of the rock is homogenous. Hackles form as distinct lines, perpendicular to ribs and are always present, even if the ribs are not visible. The shoulders are the edges of the joint face. A joint shoulder can be bounded by rock or by air if the joint formed in an isolated boulder.

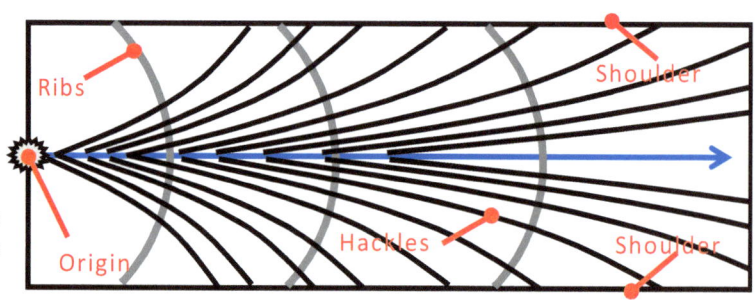

**Geologic Map**

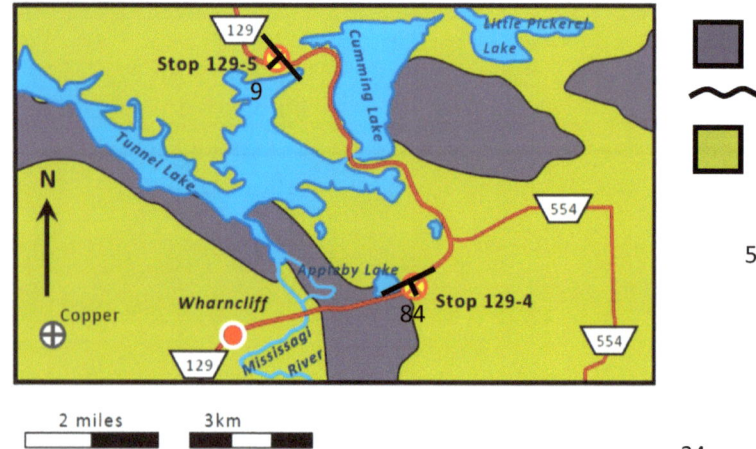

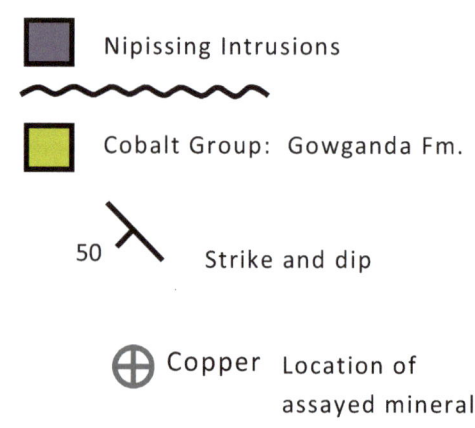

**PHOTOS:**

Outcrop, looking west. 5'10" (1.78 meter) person in front of the studied outcrop.

Joint face, looking north. The red star is the origin point of the joint. The arrow shows the propagation of the shockwave through the rock. The dotted line represents the shockwaves and are expressed in the rock as "ribs". The yellow solid lines are "hackles". Hackles are above the arrow as well. I just didn't draw them in so you can easily see them in the rock. The dashed black lines represent the shoulder.

Yellow arrows pointing to laminations in proto-quartzite, looking north. $1 piece scale.

Close-up of the proto-quartzite, looking north. $1 piece scale.

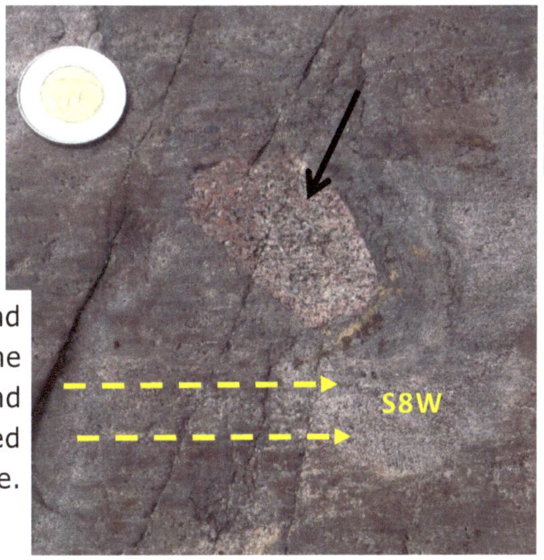

Top of outcrop, looking down and east. A granitic clast within the proto-quartzite (solid arrow) and Quaternary glacial striations (dashed arrows) point S8W. $2 piece scale.

**OUTCROP NAME:** Route 129 east side of Lafoe Creek south Outcrop  **OUTCROP DESIGNATION:** 129-6

**OUTCROP LOCATION:** GPS: 46.62985 −83.342881  ELEVATION: 975 feet (297.2 meters) above Mean Sea Level

**INFORMAL GEOLOGIC NAME:** Granite-granodiorite suite

**MAIN ROCK TYPE(S):** Granite, trachyte-syenite

**DESCRIPTION:** The outcrop is located on the east side of Route 129, just opposite of the Mississagi River about 0.76 driving miles (1.22km) south of where it meets LaFoe Creek.

The host rock at this outcrop is a typical coarse grained granite. It is likely Late Archean in age, although it has never been dated. There is the possibility that it may be Paleoproterozoic.

It is cross cut by pink to red trachyte to syenite dikes. Trachyte is the volcanic version of syenite, quartz alkali-feldspar syenite, quartz syenite, alkali-feldspar syenite, foid bearing alkali-feldspar syenite, and foid bearing syenite. Foid is an abbreviation for "feldspathoid". Feldspathoids are minerals that look like feldspar but lack quartz, such as nepheline and sodalite.

The dikes are fine grained enough in parts, to be considered volcanic, although they are intrusive and not extrusive rock. Since parts are micro-crystalline and others are not, the dikes are borderline trachyte-syenite. The dikes are not large enough or extensive enough to show up on any of the published geologic maps. They have no quartz and some likely contain foids.

**FIGURE: Geologic Map**

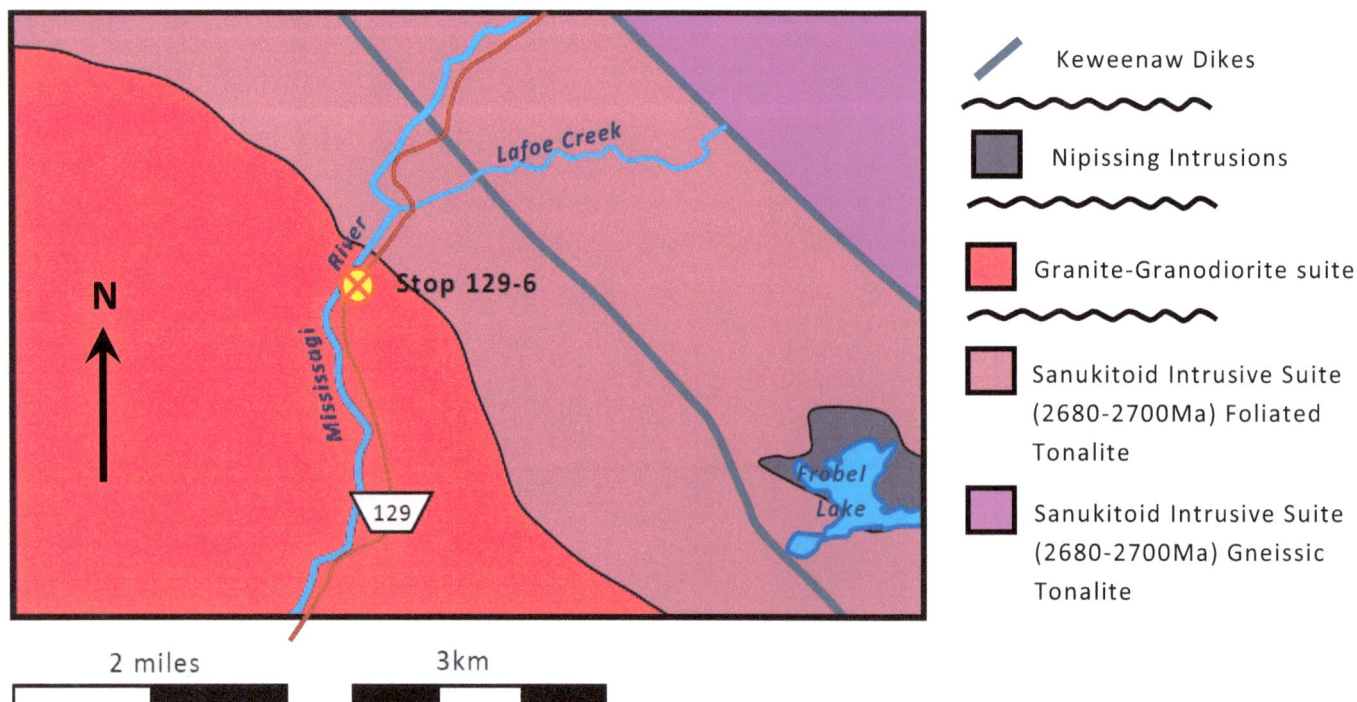

**PHOTOS:**

Outcrop, looking east-southeast. 5'10" (1.78 meter) person for scale.

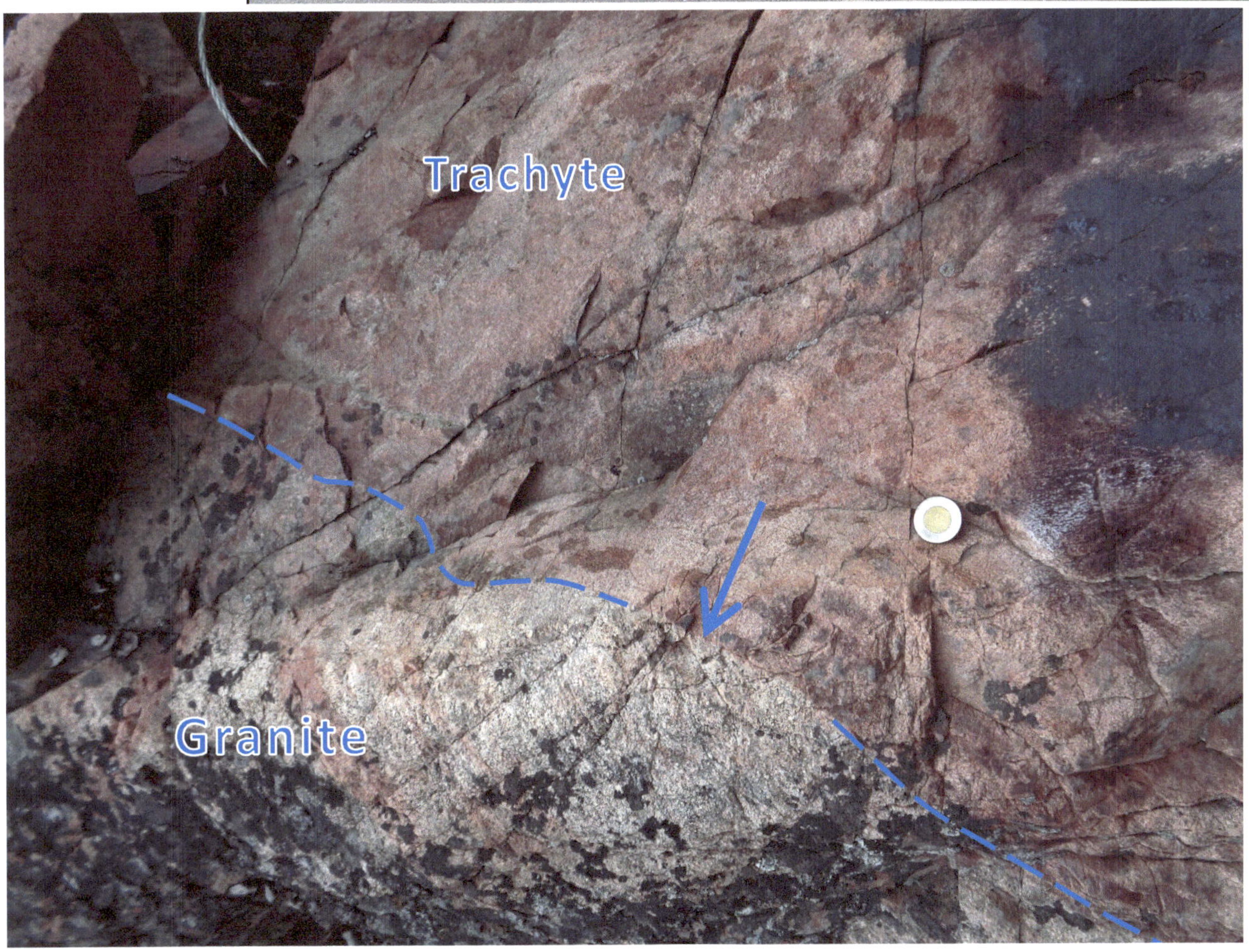

Close-up of the granite and the trachyte, looking south. The blue dashed line is the contact. The arrow is pointing to the contact (without blue dashed line) in order to show how sharp the contact is. $2 piece scale.

**OUTCROP NAME:** Route 129 east side of Mississagi River Outcrop  **OUTCROP DESIGNATION:** 129-7

**OUTCROP LOCATION:** GPS: 46.71427 –83.41797  ELEVATION: 1047 feet (319.1 meters) above Mean Sea Level

**FORMAL GEOLOGIC NAME:** Sanukitoid Intrusive Suite

**MAIN ROCK TYPE(S):** Quartz monzodiorite-diorite and minor granite, gabbro

**DESCRIPTION:** The outcrop is located on the east side of Route 129, just opposite of the Mississagi River.

At the south end of the outcrop, there is a dark greenish gray to black, coarse grained gabbro (where the GPS coordinate was taken). Its over 90% dark mafic minerals (pyroxene and olivine), with <10% white felsic minerals (plagioclase). The orientation of the gabbro intrusion was not determined. If it is similar to the other Keweenaw dikes in the area, it should trend about N50-60W.

The best place to see the host rock is along the same side of the road north about 350 feet (107 meters). The rock is part of gneissic tonalite but here it is actually a red quartz monzodiorite to granodiorite. Some isolated places the amount of quartz increases from about 10% to 20%, and the rock is closer to a granite.

A major fault, the Flack Lake Fault, occurs directly under Route 129 at this location. It is the second most significant fault in the area, second only to the Murray Fault. The exact length of the Flack Lake Fault is not known. Here you are near its western most extent. From here it travels southeast before turning almost due east towards Flack Lake. The fault is at least 40 miles (64km) long. It is suspected to have formed during the Penokean Orogeny about 1.77 billion years ago. An orogeny is a mountain building event.

**FIGURE: Geologic Map**

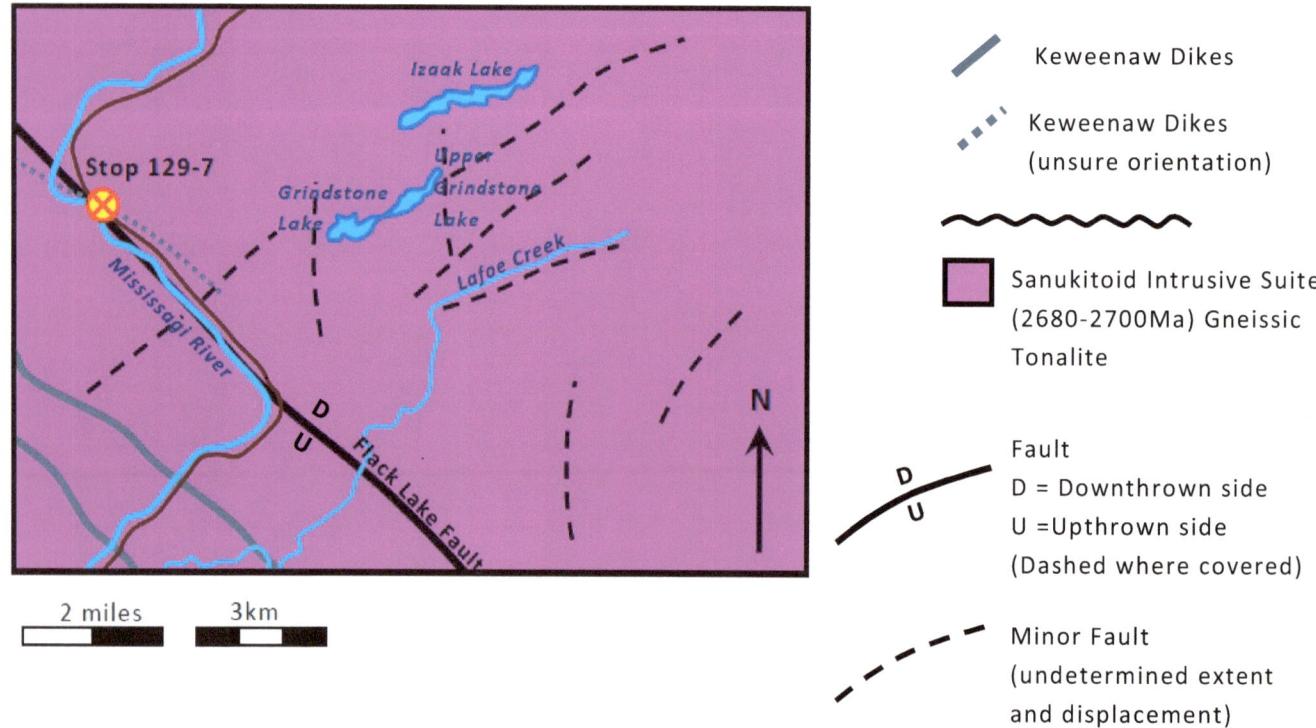

**PHOTOS:**

Outcrop, looking north-northwest. Mississagi River is on the left. No scale.

Gabbro, looking east. $1 piece scale.

North end of the outcrop, looking west-southwest. 5'10" (1.78 meter) person for scale.

Close-up of the red granite at the north end of the outcrop, looking east. $2 piece scale.

**OUTCROP NAME:** Route 129 Aubery Park entrance Outcrop  **OUTCROP DESIGNATION:** 129-8

**OUTCROP LOCATION:** GPS: 46.92201 –83.20791

**ELEVATION:** 1244 feet (379.2 meters) above Mean Sea Level

**INFORMAL GEOLOGIC NAME:** Granite-granodiorite suite

**MAIN ROCK TYPE(S):** Quartz monzonite-granite (host rock) and white quartz rich black diabase (dike)

**DESCRIPTION:** The outcrop is located on the west side of Route 129, just opposite of the first sign for "Aubrey Falls Provincial Park". The geologic map does not contain the outline of Aubrey Falls Park because of its odd shape. A detailed map can be found at ontarioparks.com/park/aubreyfalls. There is no camping but it's apparently open year round.

The host rock is the typical red granitic rocks of the area. It consists mostly of quartz monzonite to granite. Primary quartz makes up <25% of the rock. Typically around 5% to 10%. The rock here is very coarse so it is a good place to see the striations present on the red plagioclase.

There is one known Keweenaw dike about 1 mile (0.6km) to the southeast. At the outcrop there are two black diabase dikes. The biggest is about 6 feet (1.83 meters) in width and contains highly folded white quartz. The with quartz also penetrates the host rock as sharp knife blade like intrusions. The quartz shows complicated folding and fabric. Due to time constraints, I did not study it in detail. The origin of the quartz is likely postdates even the diabase (which is assumed to be Keweenaw in age based on its trend) and is likely due to hydrothermal activity. The complex internal structure could be due to folding that occurred if the dike follows a fault that remained active after the quartz was deposited. There is a smaller similar dike about 150 feet (46 meters) to the north along the outcrop.

**FIGURE: Geologic Map**

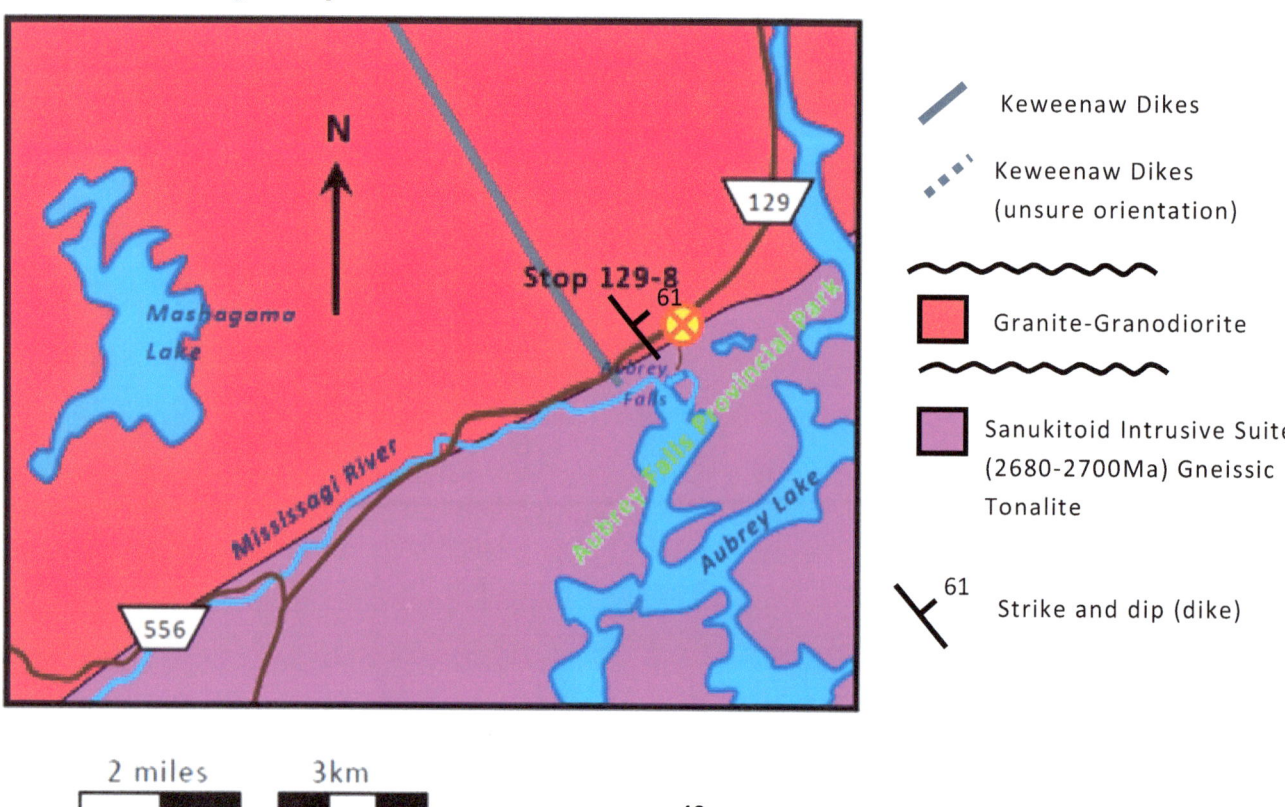

## PHOTOS:

Outcrop, looking west-northwest. 5'10" (1.78 meter) person for scale.

Dark diabase dike with white quartz trends N40W61NE, looking northwest. 6" (15.2cm) orange pencil scale (red circle).

Close-up of the contact (yellow dashed line) between the dike and the host rock, looking northwest. The black arrows are pointing to the knife blade like quartz intrusions into the host rock. $2 piece scale (red circle).

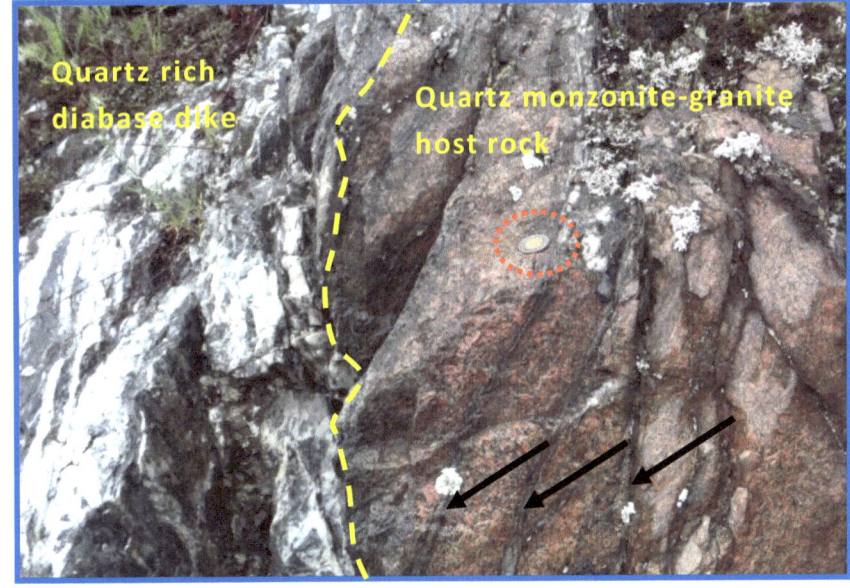

Quartz rich diabase dike

Quartz monzonite-granite host rock

Close-up of biotite rich (black parts) quartz monzonite to granite (red rock) of the host rock, looking northwest. $2 piece scale.

**OUTCROP NAME:** Route 129 opposite Flame Lake north Outcrop  **OUTCROP DESIGNATION:** 129-9

**OUTCROP LOCATION:** GPS: 47.28393 −83.20694

**ELEVATION:** 1527 feet (465.4 meters) above Mean Sea Level

**INFORMAL GEOLOGIC NAME:** Granite-Granodiorite Suite

**MAIN ROCK TYPE(S):** Breccia in quartz vein, hornblende-biotite red granite

**DESCRIPTION:** The outcrop is located on the west side of Route 129, about 0.32 miles (0.51km) north of the road that leads to Flame Lake Lodge.

The host rock is a typical course grained hornblende-biotite granite that is close to the quartz monzonite line. It's not very impressive.

What is impressive is the brecciated quartz vein. It appears to follow the dominant fracture regime and trends N45E81NW. The red angular breccia varies from pebble to boulder size and consists of the host rock. The breccia sits inside a mass of white quartz. The white quartz contains a disseminated very fine grained black metallic mineral, along with mica and hornblende. There are also occasional pink phenocrysts, that are likely feldspar. On the north side, the contact between the vein and host rock is sharp and unaltered. This is a strong indication that the quartz is not magmatic in nature but rather hydrothermal. On the south side of the vein, the contacts are sharp but small quartz dikelets (small veins and dikes) propagate through the host rock past the main breccia vein. The breccia vein has an apparent width of 14 feet (4.3 meters) along the outcrop surface. It's actual width (or thickness) is 10 feet (3.0 meters).

**FIGURE: Geologic Map**

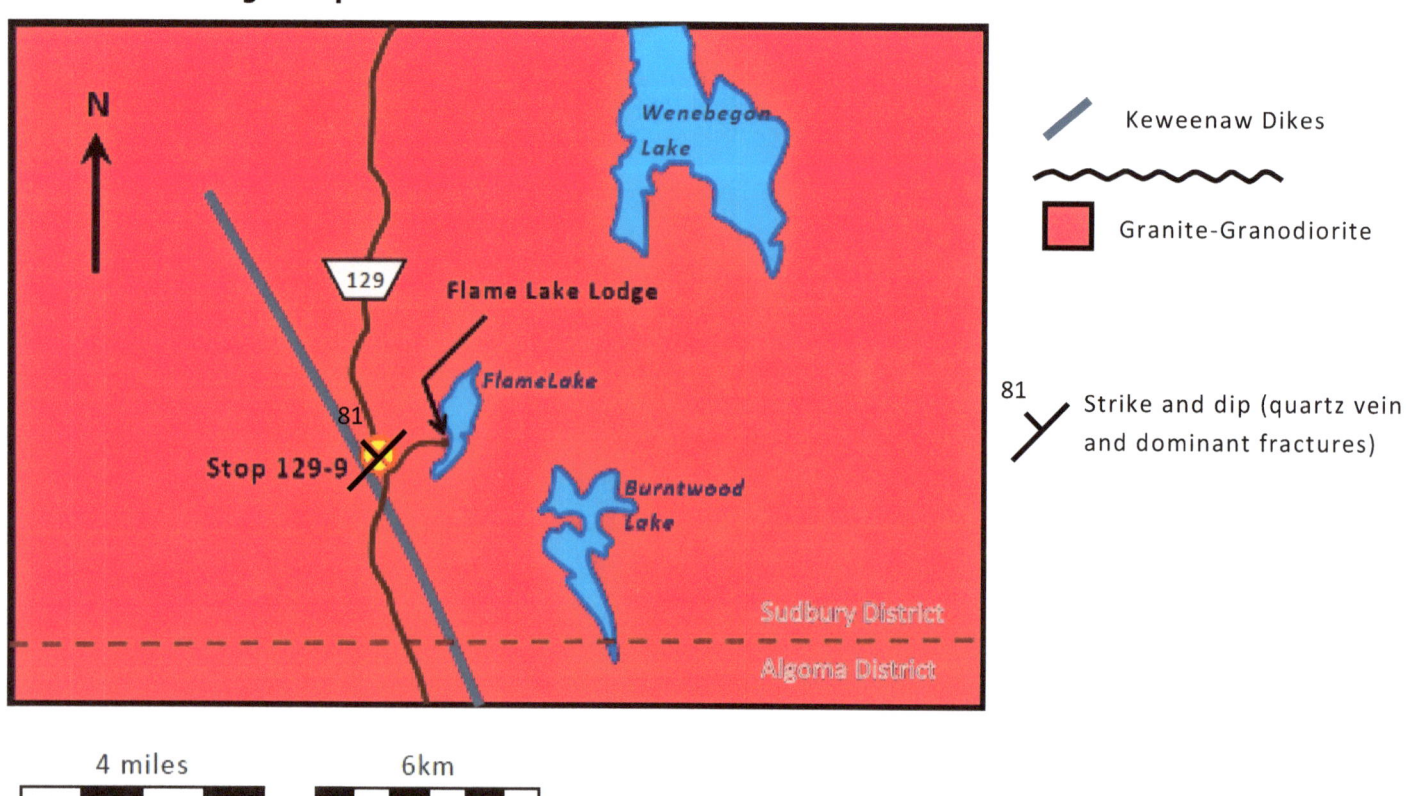

**PHOTOS:**

Outcrop, looking west. 5'10" (1.78 meter) person for scale.

Close-up of breccia vein, looking west. Arrow pointing to $2 piece scale.

Contact of the breccia and quartz vein with the host rock, looking west. The black arrows point to the contact. $2 piece scale sits on the red host rock.

**OUTCROP NAME:** Route 129 between Nemi and Pike Lakes Outcrop  **OUTCROP DESIGNATION:** 129-10

**OUTCROP LOCATION:** GPS: 47.35059 –83.21021  ELEVATION: 1480feet (451.1 meters) above Mean Sea Level

**INFORMAL GEOLOGIC NAME:** Granite-Granodiorite Suite

**MAIN ROCK TYPE(S):** Biotite granite with biotite clasts

**DESCRIPTION:** This is a massive outcrop is located on both sides of Route 129, about halfway between Nemi and Pike Lakes. Both sides of the outcrop were recently cut out of the rock and the rock is similar on both sides of the road.

Starting at this outcrop and heading north and west, the red Archean rocks will become rarer and rarer. This outcrop appears on Map P-674 (1971) and is the most accurate description of the outcrop, listing it as a "massive to foliated biotite and biotite-hornblende granite". Newer maps such as M-2543 (1991) list it as "gneissic tonalite". This rock is neither gneissic nor tonalite.

Although this outcrop is listed as part of the Granite-Granodiorite Suite, it likely is not. The host rock is finer grained and contains far more gray and white feldspar than the rocks to the south. It is a massive granite with isolated pink pegmatitic parts of alkali-feldspar. It also contains abundant biotite, which is the black mineral that makes up about 10% to 15% of the rock. The rock is likely a small stock within the Granite-Granodiorite Suite. It has never been dated, but based on field relationships it is likely Neoarchean.

The most interesting thing about this outcrop are the isolated but relatively large clasts of black concentrated biotite and minor hornblende, that forms "wispy" patterns within the homogenous granite. I call this pattern "smoke rock" as it forms random smoke like patterns within the host rock. They likely formed as cooled but plastic minerals were churned about in a still thick and viscous magma.

**FIGURE: Geologic Map**

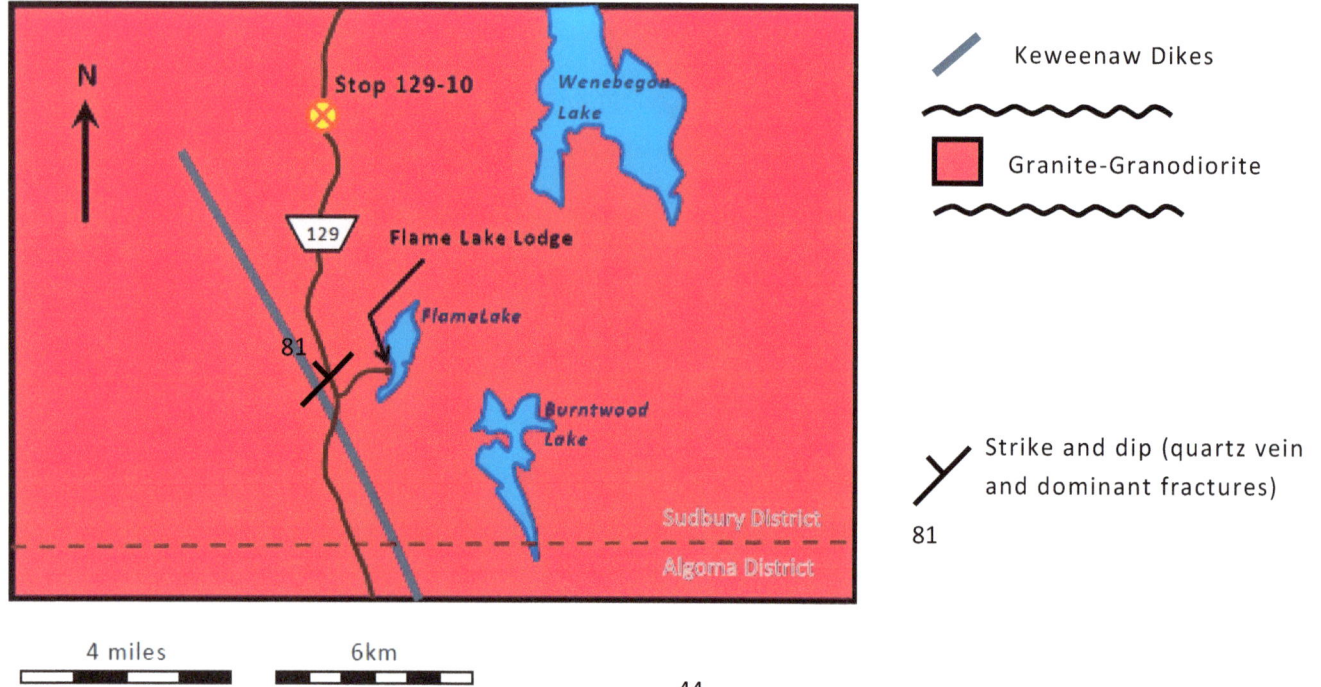

**PHOTOS:**

Outcrop, looking north. This outcrop is a new road cut, looking north. 5'10" (1.78 meter) person for scale.

Close-up of the host rock, looking east. $2 piece scale.

"Smoke rock" clast, looking east

Pink pegmatitic part of alkali-feldspar, looking east. $2 piece scale.

**OUTCROP NAME:** Route 101 1.4 mile west of Route 129 Outcrop   **OUTCROP DESIGNATION:** 101-12

**OUTCROP LOCATION:** GPS: 47.75743 –83.41492   ELEVATION: 1477feet (450.2 meters) above Mean Sea Level

**INFORMAL GEOLOGIC NAME:** Granititc migmatite and gneiss

**MAIN ROCK TYPE(S):** Gneissic hornblende-biotite and granodiorite (migmatite)

**DESCRIPTION:** The outcrop is about 1.40 driving miles (2.25km) west of the Route 129-101 junction on both sides of Route 101.

There are low outcrops on both sides of the road. Both express a highly foliated gneissic pattern cause by dark blue to black concentrations of biotite and minor hornblende. The host rock (white on the south side and pink on the north) is undeformed. The masses of biotite were likely incorporated into a magma, but not melted themselves, as the feldspar eventually crystalized around it. The evidence for this is in the circular pattern of the biotite. The pattern is conducive of convection cells.

Both sides of the outcrop cooled from the same magma, but there is a slight difference. The north outcrop contains almost all white plagioclase for its feldspars. As where the south outcrop contains pinkish feldspar which has a slightly higher alkali-feldspar content than the north side. The south outcrop feldspars are also coarser grained. What this likely represents is the south outcrop cooled second after the north outcrop from the same magma body. The south outcrop would have been closer to the center of the magma chamber than the north outcrop. In other words, the magma to the south cooled slower, allowing for a cation exchange from plagioclase to a-k feldspar, according to Bowen's reaction series.

I have assigned the rocks in the local area to Mesoarchean. This is based strictly off observed field relationships. To my knowledge, the rocks themselves have never been dated, and maybe older or younger than Mesoarchean.

**FIGURE: Geologic Map**

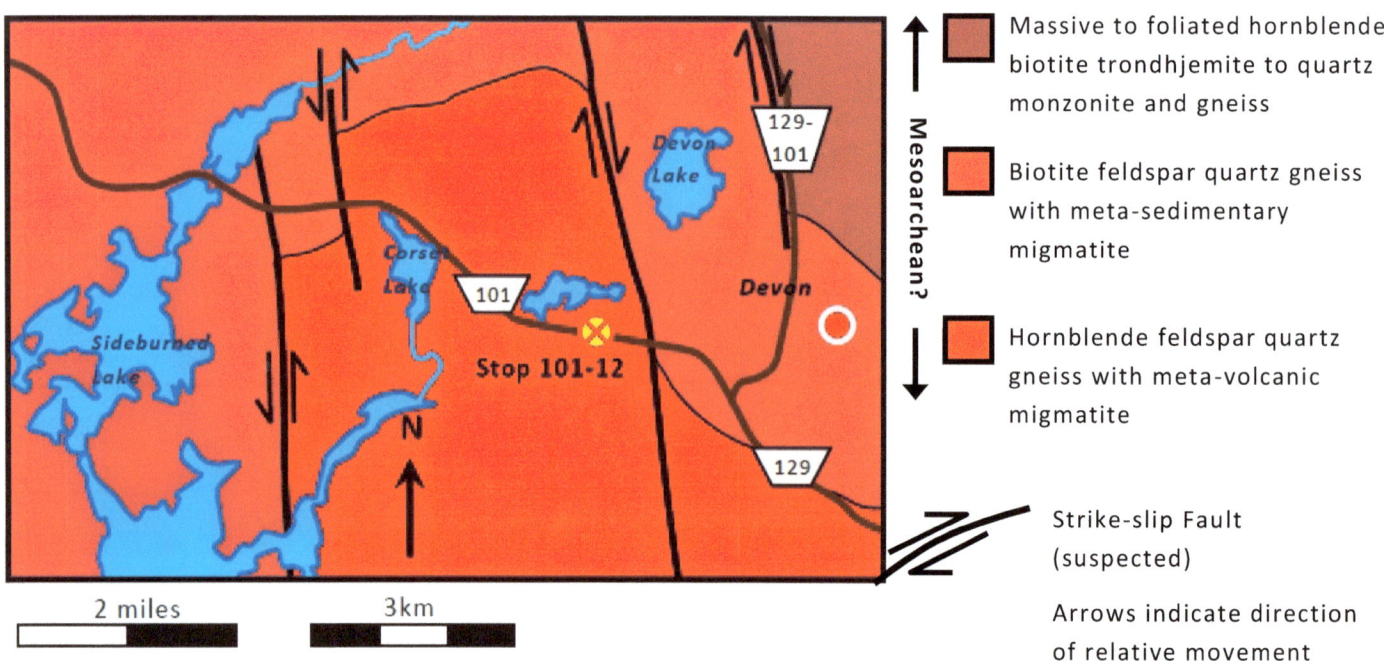

**PHOTOS:**

Outcrop, looking south. 5'10" (1.78 meter) person for scale. South outcrop.

Dark diabase dike with white quartz trends N40W61NE, looking south. 6" (15.2cm) orange pencil scale. South outcrop.

Outcrop, looking north. 5'10" (1.78 meter) person for scale. North outcrop.

The biotite pattern hasn't changed. However, the feldspar is coarser grained and somewhat pegmatitic on this side of Route 101, looking north. $2 piece scale. North outcrop. The inset is a zoom-in of the rock. The red arrow points to a large dark blue flake of biotite.

**OUTCROP NAME:** Route 101 west of kilometer marker 123 Outcrop   **OUTCROP DESIGNATION:** 101-11

**OUTCROP LOCATION:** GPS: 47.77360 −83.48022   ELEVATION: 1511feet (460.6 meters) above Mean Sea Level

**FORMAL GEOLOGIC NAME:** Wawa Stock

**MAIN ROCK TYPE(S):** Gneissic hornblende-biotite (migmatite), with syenite dike

**DESCRIPTION:**

The outcrop is about 0.48 miles (0.77km) east of the bridge over Sideburned Lake. It is extensive and on both sides of the road. I studied the part where a red dike was present (at the above GPS coordinates).

The host rock is extremely gneissic and the parent rock is not easily identifiable. What can be said about it is the black foliations are mostly biotite but amphibole (hornblende) is also present. The feldspars present are white to gray and are mostly plagioclase. Quartz is minimal. It is possible that the parent rock of the gneiss was sedimentary as it tends to be very planar and layered.

The most interesting thing about this outcrop is the red dike that crosscuts the gneiss. Steno's stratigraphic law of "lateral continuity" says, even though sections of rock may be eroded, if deposited in a similar environment the rocks are the same units and will be continuous until a change in structure occurs. The red dike on both sides of Route 101 line up and have the exact same strike and dip and are the same thickness at about $10.2\pm0.5$ inches ($25.9\pm1.3$cm). The dike is almost entirely red alkali-feldspar with about 5% primary quartz and isolated black hornblende. The abundance of alkali-feldspar tells us the dike intruded the rock much later, likely post Archean. It hasn't been dated, so its actual age is not known. The contact of the dike with the host rock is sharp. The lower contact shows no alteration. The upper contact has alteration and the feldspars right above it tend to be pink instead of the white that is typical in the host rock. This pink color can be explained by simple heat transfer from the dike as it cooled to the cooler gneiss above, as opposed to the warmer gneiss beneath. The red dike forms at the opposite angle of the fractures in the rock and itself is fractured in the north outcrop, forming an "x" shape.

A large unnamed north-south trending fault has been mapped very close by. The studied outcrop shows no sign of faulting.

**FIGURE: Geologic Map**

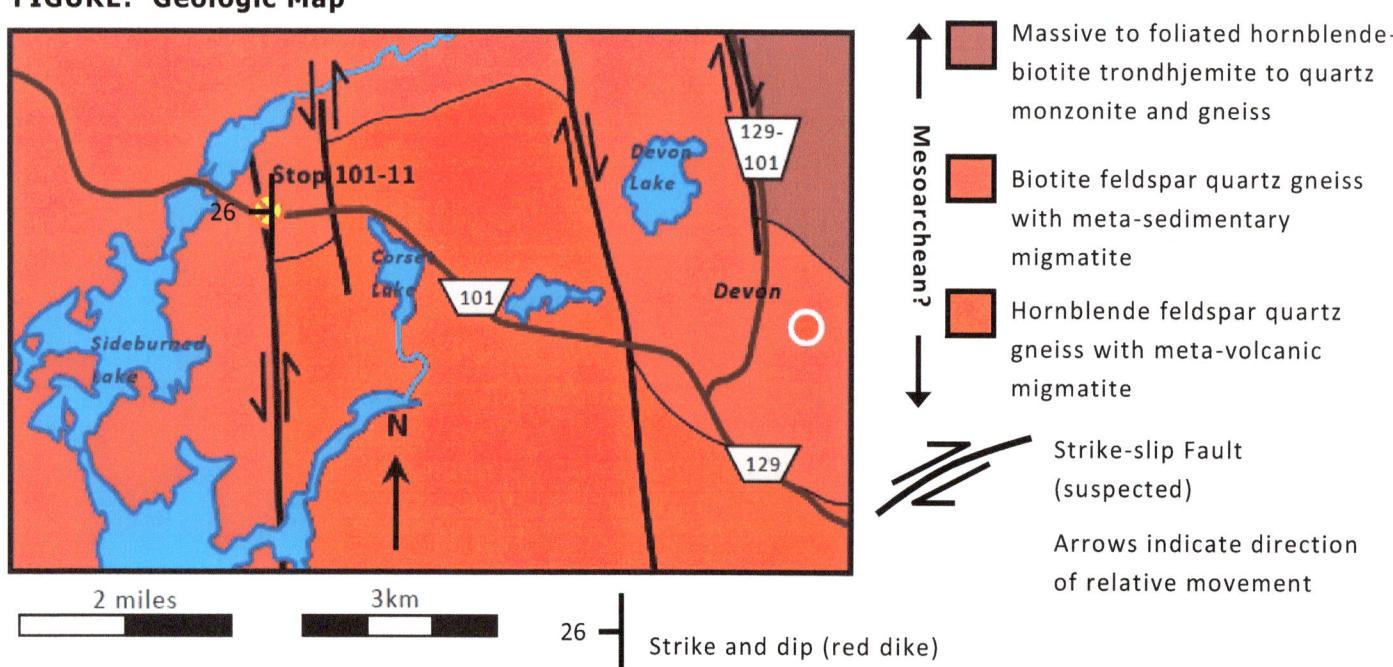

**PHOTOS:**

Outcrop, looking north. 5'10" (1.78 meter) person for scale.

Outcrop, looking south. 5'10" (1.78 meter) person for scale.

Person is pointing to a red dike in both photos. It is actually the same dike. When Route 101 was built, the middle of the outcrop was removed. This type of correlation can be done because of the "Law of lateral continuity".

Red quartz syenite dike, looking north. The dike follows existing fractures and has sharp contacts with the host rock (arrows). The top contact shows an altered zone, the bottom does not. This contact pattern holds on both sides of the road. 6" (15.2cm) orange pencil scale.

Red quartz syenite dike, looking south. The host rock above and below the dike follows the trend of the dike as where in the photo above, the dike crosscuts the gneiss foliation in the host rock. This could mean something or it may be coincidence. 6" (15.2cm) orange pencil scale on top of the dike.

**OUTCROP NAME:** Route 101 kilometer marker 119 Outcrops   **OUTCROP DESIGNATION:** 101-10

**OUTCROP LOCATION:** GPS: 47.78159 −83.51917    ELEVATION: 1537 feet (468.5 meters) above Mean Sea Level

**FORMAL GEOLOGIC NAME:** Wawa Stock

**MAIN ROCK TYPE(S):** Biotitic (mafic) gneiss, monzonite-quartz monzonite, basalt dike

## DESCRIPTION:

The studied outcrop is about 0.34 driving miles (0.55km) of the northern tip of Highbrush Lake. It occurs as sporadic outcrops but only the above GPS was studied.

The host rock consists of an igneous red monzonite-quartz monzonite and a black biotite gneiss, making it a migmatite. The monzonites intruded the gneiss and were likely partially (if not entirely) derived from it as this outcrop experienced partial melting deep underground. It is impossible to determine the parent rock from the gneiss since the rock has been altered so much.

There is a large basal dike that intrudes the outcrop. The dike where in contact with the host rock is about 3.90 feet (1.19 meters) to 4.25 feet (1.30 meters) thick. Its orientation changes from N15W67NE at its north end, to N3W35NE in the middle (mapped below), and N7W39NE at the south end. This is exactly opposite to the syenite dike at stop 101-11. As if the changing of orientation weren't enough, the dike sits within a rounded basalt mass near the top of the outcrop. The nature of this mass isn't clear. It is unusual for a basaltic dike to form such a chamber before continuing to the surface. It is possible the rounded mass is not part of the dike but actually an older intrusion. This is possible since the rounded mass contains monzonite stringers within it. Either way, it is clear that the host rock was still deeply buried and still warm when the basalt dike intruded it. This can be deduced because the dike cuts through the host rock unevenly and is in sharp contact with it. The extent of the dike is not known. It only on one geologic map (Map P.423, 1967) and is labeled as diabase. Its age is also unknown, but basaltic dikes like this are usually Keweenaw (~1.1 billion years) in age. Although it could date back to the time of the Matachewan dike swarm, which has been dated at 2633±75 million years ago, which would make the Wawa Stock older than that.

This outcrop sits on a structure called the "Highbrush Lake Dome", a series of local low lying domes.

**FIGURE: Geologic Map**

**PHOTOS:** Composite photo of the outcrop, looking east. 5'10" (1.78 meter) person for scale. The host rock is a red monzonite-quartz monzonite, with black biotitic gneiss. It is intruded by an irregular shaped dark gray basalt dike (within yellow dashed lines). Although the basalt forms an irregular rounded mass near the top of the outcrop, it continued up through the host rock and the rounded mass as a typical dike (yellow dotted lines) before completely cooling, assuming the rounded mass is from the same magma as the dike.

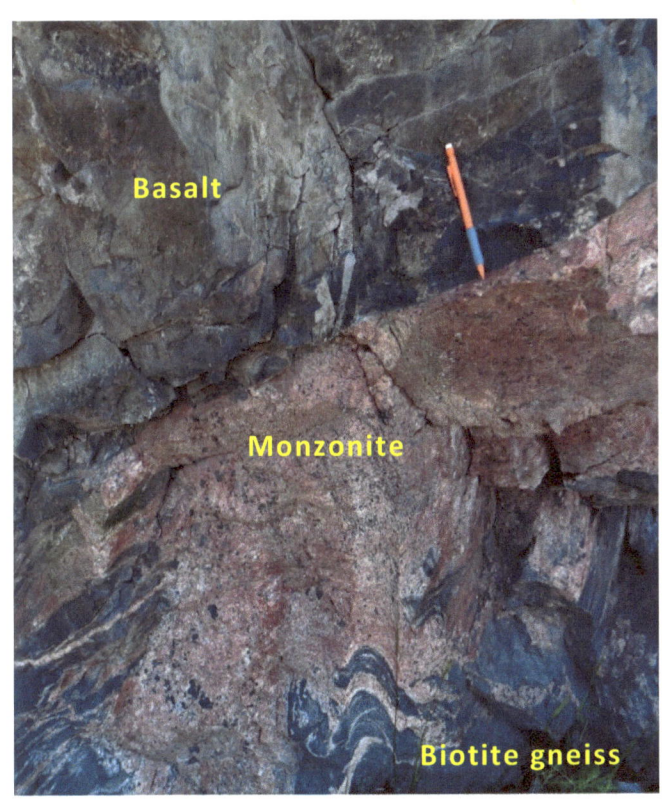

The sharp contact between the basalt dike and the host rock is being pointed to by the 6" (15.2cm) orange pencil. The host rock consists of monzonite and biotite gneiss, making it a migmatite, looking east.

Here the monzonite both crosscuts the gneiss (yellow arrows) and flows in between it (red arrow). 6" (15.2cm) orange pencil scale.

**OUTCROP NAME:** Route 101 kilometer marker 94 east Outcrop  **OUTCROP DESIGNATION:** 101-9

**OUTCROP LOCATION:** GPS: 47.85711 –83.80138  ELEVATION: 1514 feet (461.5 meters) above Mean Sea Level

**FORMAL GEOLOGIC NAME:** Wawa Stock

**MAIN ROCK TYPE(S):** Gneiss, tonalite/trondhjemite, basalt-gabbro

**DESCRIPTION:** The studied outcrop is on both sides of the road within "The Shoals Provincial Park". The website is: ontarioparks.com/park/theshoals. It is about 0.35 miles (0.56km) west-northwest of the small bridge crossing the Grazing River.

There are three main rock types exposed on both sides of the road.

On the north outcrop you see the Archean white biotite gneiss host rock. There is also a massive black dike. The dike is micro crystalline to medium grained, so it is a basalt-gabbro. The age of the dike is not known. However, there are some small drilled cores that have been removed from the base. In 2015 the Keweenaw dikes were extensively sampled for paleomagnetism data, as part of a dissertation done to study the Neoproterozoic geomagnetic field. So the dike here may be Keweenaw. The width of the dike along the outcrop is over 100 feet (30 meters). Its orientation is N35W70NE, making its actual width about 56 feet (17.1 meters).

Across the road, the gneiss is much better exposed and more biotite rich. There are also white pegmatitic tonalite veins. They tend to be long and straight. Almost all the feldspar in them is plagioclase, which is evident by the abundant striations. That would make the veins technically trondhjemite. It is a bit odd for such thin veins to have such large crystals. It is likely they formed from melting the gneiss as very hot gasses and fluids moved between the fractures while the gneiss was still deeply buried.

**FIGURE: Geologic Map**

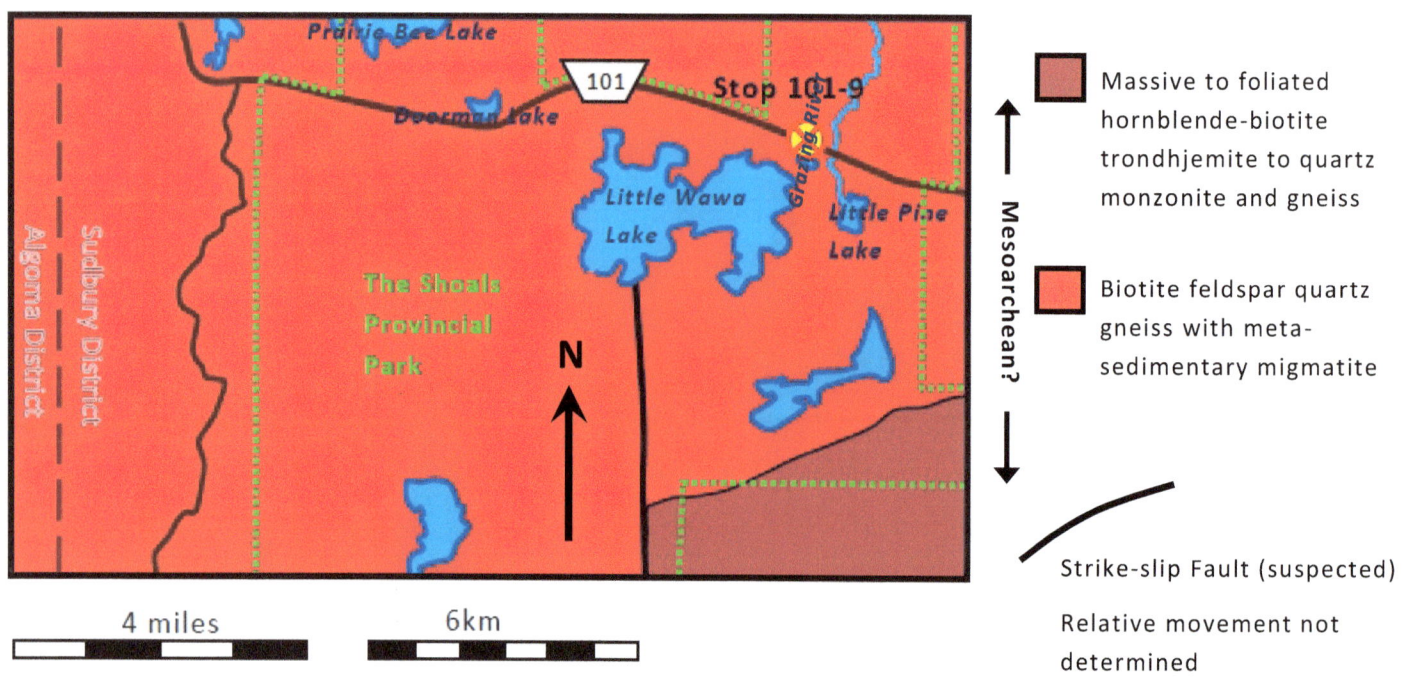

**PHOTOS:**

Outcrop, looking northwest. 5'10" (1.78 meter) person for scale, near the center. The white rock is the host rock and is a biotite gneiss. The dark rock is basalt-gabbro dike.

Contact between the dike and the host rock, looking north. 6" (15.2cm) orange pencil scale.

Biotite rich part of the host rock, looking south. 6" (15.2cm) orange pencil scale (circle). Arrow points to pegmatitic tonalite dike.

Close-up of the pegmatitic white plagioclase crystals in a vein, looking south. 6" (15.2cm) orange pencil scale. There is some primary quartz, and some black biotite in the rock. Technically, the rock is a pegmatitic trondhjemite.

**OUTCROP NAME:** Route 101 just north of kilometer marker 83 Outcrop  **OUTCROP DESIGNATION:** 101-8

**OUTCROP LOCATION:** GPS: 47.86964 −83.94509   **ELEVATION:** 1570 feet (478.5 meters) above Mean Sea Level

**FORMAL GEOLOGIC NAME:** Wawa Stock

**MAIN ROCK TYPE(S):** Gneiss, monzonite

**DESCRIPTION:** The studied outcrop is on both sides of the road about 0.3 miles (0.48km) due west of Goodwin Lake.

Both sides of the outcrop contain similar rocks. The outcrop on the west side of the road is a highly foliated gray gneiss. The white parts are feldspar and quartz. The black parts are mostly biotite. Based on field data only, it isn't possible to determine whether the parent rock of the gneiss was igneous or sedimentary. The internal deformation is just way too chaotic.

The entire outcrop is cross cut by pink to red monzonite dikes that have an apparent dip of about 29°/N9W. These dikes are also present on the east outcrop. They are thinner, less numerous, and weathering conceals them better. The east monzonite dikes trend about 25°/N10W. Since these two apparent dips are similar, it is likely very close to true strike and dip for the dikes. We can't say for sure because none of the dikes can be confidently correlated to both sides of the road.

The gneiss host rock on both sides of the road displays pseudo-random, very chaotic, and disconnected folds. These type of folds are called ptygmatic folds. On occasion, there is some discernable structure, which is more prominent on the east outcrop. There will often be small "isoclinal recumbent folds", that are in between small "shear zones". Shear zones are very complicated geologic structures. For purposes of this guidebook, shear zones are best described as a type of fault where ductile (or fluid) deformation is favored over brittle (breaking) deformation, as seen in typical faults. This mini shear zone experienced only ductile deformation, indicating the rock was very hot and under a lot of pressure, but not enough to entirely melt it.

**FIGURE:** Geologic Map

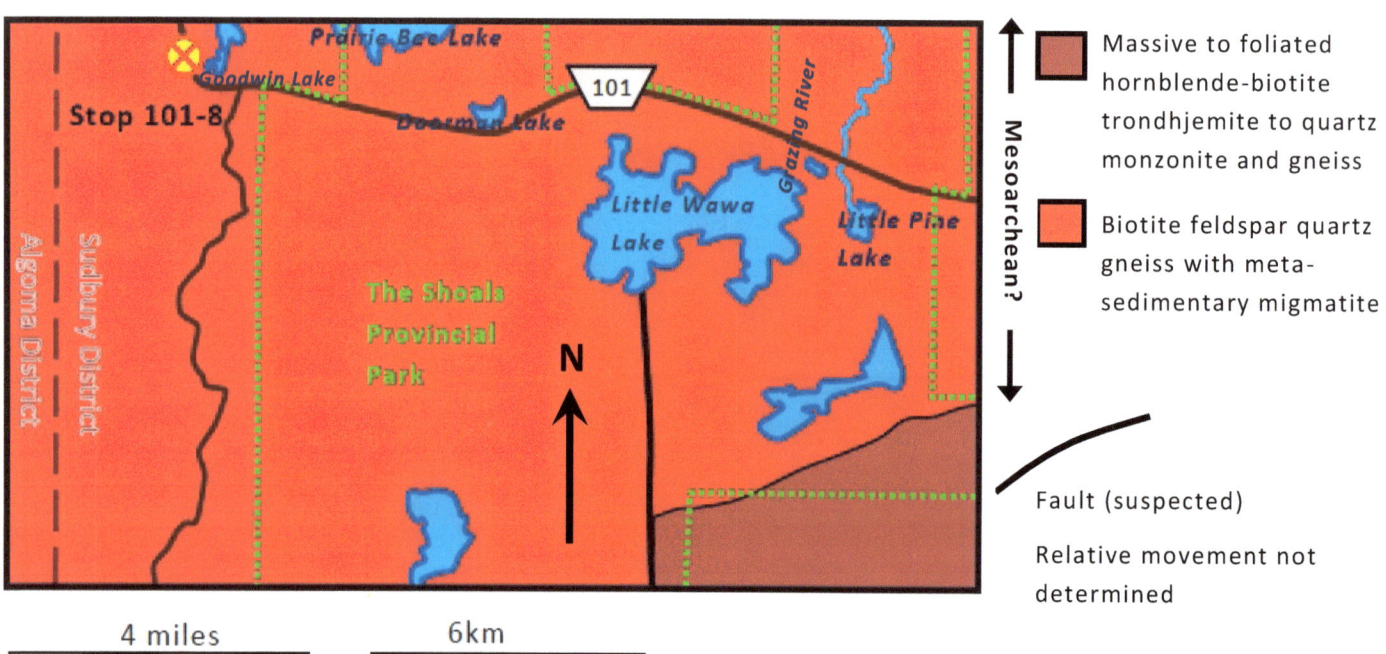

**PHOTOS:**

Outcrop, looking west. 5'10" (1.78 meter) person for scale.

Outcrop, looking east. 5'10" (1.78 meter) person for scale.

Close-up of monzonite dike, looking west. Arrow is pointing to $1 piece scale.

Gneiss showing recumbent folding (see inset photo in yellow box for details). Yellow arrows are pointing to the barely visible monzonite dikes (apparent dip 25°/N10W), looking east. 6" (15.2cm) orange pencil scale.

In this inset the solid yellow lines indicate "shear zones" within the gneiss and the yellow arrows indicate relative movement. The relative movement of the shear zones can be determined by the offset of the black mafic rock. 1a and 1b used to be one piece. Same with 2a and 2b. These shear zones caused an isoclinal recumbent fold to form (black lineation in the natural rock through the red lines). This type of fold is a sideways symmetrical fold where the fold axis (red lines) are parallel to one another.

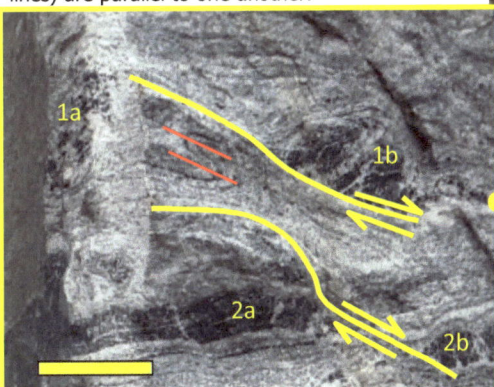

**OUTCROP NAME:** Route 101 on Quill-Nadjiwon Township line Outcrop  **OUTCROP DESIGNATION:** 101-7

**OUTCROP LOCATION:** GPS: 47.93364 –84.11092  ELEVATION: 1497 feet (456.3 meters) above Mean Sea Level

**FORMAL GEOLOGIC NAME:** Wawa Stock

**MAIN ROCK TYPE(S):** Gneiss to migmatite, tonalite, basalt

**DESCRIPTION:**

The outcrop about 0.73 driving miles (1.12km) southeast of the Route 101/651 T-junction.

There are three main rock types present here and they are likely the same age as the rocks at stops 101-8 to 101-12. The host rock is a felsic gneiss. What that means is the vast majority of the rock is mafic minerals such as biotite (both golden brown and dark blue at this outcrop as small to large flakes) and fine weakly crystalline hornblende (black blade-like mineral). Although the foliation appears random, it is less so than at Stop 101-8 and actually has a preferred average trend of about N27E87SE. This stop and stops 101-8 to 101-10 may actually be the same metamorphic body. Its just this stop contains very little migmatite. There are some places that were clearly melted, but they are isolated. Possibly because it's closer to the edge of the intrusion where no melting took place. That doesn't mean all was quiet.

There are two intrusions with very different mineralogy and texture. The most interesting is the "brecciated tonalite dike". I actually took a piece of this home and counted the minerals to figure out what it was. The white clasts are white plagioclase that have been partially melted and altered because some have pink edges. There is primary gray quartz in it. The clasts exist in a nearly black groundmass. The plagioclase was likely ripped up from some great depth as this dike mobilized to the surface in what is now the groundmass. It trends nearly due north-south. It has not been dated, nor does it really belong to any known dike swarm. So, all we can say about its age is that it's younger than the gneiss host rock.

The other is a small highly eroded black "basalt dike" at the north end of the outcrop. The dike weathers far faster than the surrounding gneiss. Unlike the gneiss the basalt is all mafic minerals, which according to Bowen's Reaction Series, weather first at surface conditions. The basalt dike is likely Keweenaw in age and probably followed existing fractures. From this point to Lake Superior, basalt dikes very commonly fill old fractures and faults in the Archean coarse grained igneous and meta-igneous rocks.

**FIGURE: Geologic Map**

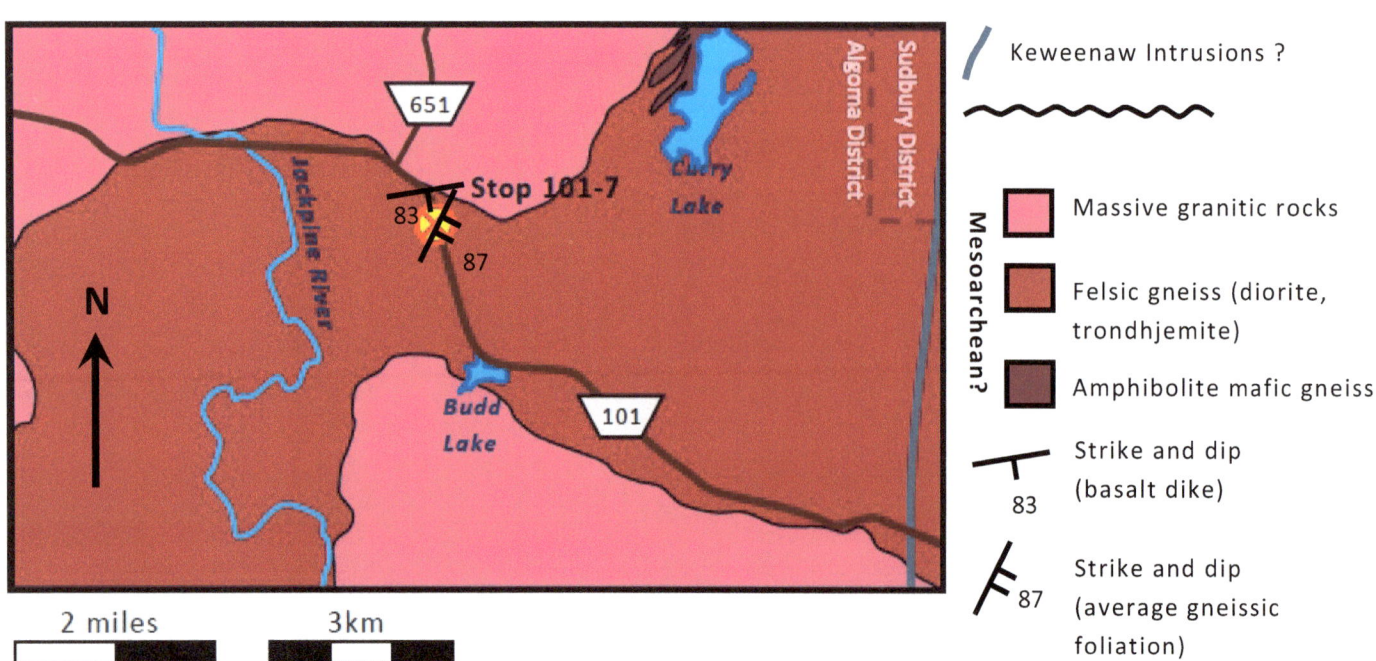

## PHOTOS:

Zoom-in of sign in photo to the right.

Outcrop, looking west-southwest. 5'10" (1.78 meter) person for scale. Yellow arrow pointing to eroded basalt dike.

Close-up of the brecciated tonalite dike, looking east. $2 piece scale. To me, this is perhaps the most beautiful dike in this guidebook. It has a complicated history and it's a great place to see striations on large partially melted plagioclase crystals.

Black basalt dike with fractures sets at nearly 90° from one another. Some fractures are parallel to the edge of the intrusion, others are perpendicular. It trends N81E83SE, looking west. 6" (15.2cm) orange pencil scale (left center).

# References

Ayer, J.A., Goutier, J., et.al, 2010, *Tectonic and Metallogenic Evolution of the Abitibi and Wawa Subprovinces*, Ontario Geological Survey (ONGS), Open File Report 6260, p.3-1 to 3-6.

Baumann, S.D.J., Cory, A.B., et. al, 2017, *Precambrian Geologic Events in the Mid-Continent of North America*, Midwest Institute of Geosciences and Engineering (MIGE), Publication: G-012011-1J

Baumann, S.D.J., 2018, *Stratigraphic Column and Ages of the Huronian Supergroup and the Chocolay Group*, MIGE, Publication: G-082017-1B

Bennett, G., 2006, *The Huronian Supergroup between Sault Ste. Marie and Elliot Lake*, Institute on Lake Superior Geology (ILSG), Field Trip Guidebook, v.52, Part 4

Bornhorst, T.J., Hanson, M.J., 2017, *Wawa, Ontario, May 8-12, 2017*, ILSG, Field Trip Guidebook, v.63, Part 2

Bowen, N.L., 1922, *The Reaction Principle in Petrogenesis*, Journal of Geology, v.XXX, no.3, p.177-198

Canfield, D.E., Ngombi-Pemba, L., Hammarlund, E.U., et al., 2013, *Oxygen dynamics in the aftermath of the Great Oxygenation of Earth's Atmosphere*, Proceedings of the National Academy of Sciences (PNAS), v.110, no.42, p.16736-16741

Colivine, 2011, *Bowen's Reaction Series*, PNG diagram, public domain, Wikipedia

Davidson, A., 1995, *A Review of the Grenville orogen in its North American type area*, AGSO Journal of Australian Geology & Geophysics, v.16, no.1/2, p.3-24

Geissman, J.W., Bowing, S.A., Babcock, L.E. (compilers), 2018, *GSA Geologic Time Scale, V.5.0*, The Geological Society of America (GSA)

Jackson, S.L., 2001, *On the Structural Geology of the Southern Province between Sault Ste. Marie and Espanola, Ontario*, ONGS, Open File Report 5995

Mustard, P.S., 1985, *Sedimentology of the Lower Gowganda Formation, Coleman Member (Early Proterozoic) at Cobalt, Ontario*, National Library of Canada, Canadian Theses Service, Department of Geology, Carleton University in Ottawa, Ontario

# References

Nutman, A.P., Bennett, V.C., Friend, C.R.L., et. al, 2016, *Rapid emergence of life shown by discovery of 3,700 million year old microbial structures*, Nature vol.537, p.535-537.

Piispa, E.J., 2015, *Precambrian geomagnetic field and geodynamics recorded selected mafic dike swarms in India and North America*, Ph.D. dissertation, Michigan Technological University, Houghton, Michigan, p.166.

Robertson, J.A., 1967, *Geology of the Spragge Area*, Ontario Department of Mines, Geology Branch, Open File Report 5010, 1966 Project 61-6

Rousell, D.H., Brown, G.H. (Editors), 2009, *A Field Guide to the Geology of Sudbury Ontario*, Ontario Geological Survey (ONGS), Open File Report 6243

Thompson, A.B., England, P.C., 1984, *Pressure-Temperature-Time Paths of Regional Metamorphism II. Their Inference and Interpretation using Mineral Assemblages in Metamorphic Rocks*, Journal of Petrology, v.25, Part 4, p.929-955

Thurston, P.C., Siragusa, G.M., Sage, R.P., 1977, *Geology of the Chapleau Area, Districts of Algoma, Sudbury, and Cochrane*, Ontario Division of Mines, Geoscience Report 157

Wentworth, C.K., 1922, *A scale of grade and class terms for clastic sediments*, The Journal of Geology, v.30, no.5, p.377-392

Winter, J.D., 2001, *An Introduction to Igneous and Metamorphic Petrology*, Text Book, published by Prentice Hall, ISBN: 0-13-240342-01

Zolnai, A.I., Price, R.A., Helmstaedt, H., 1983, *Regional cross section of the Southern Province adjacent to Lake Huron, Ontario: implications for the tectonic significance of the Murray Fault Zone*, Canadian Journal of Earth Sciences, DOI: 10.1139/e84-048

# References

**Geologic Maps produced by the Ontario Geological Survey and Ministry of Mines:**

Map 2108; 1967: Sault Ste. Marie-Elliot Lake Sheet: Scale =1 :253,440

Map 2149; 1979: Sault Ste. Marie-Elliot Lake : Scale = 1:253,440

Map 2116; 1967: Chapleau-Foleyet Sheet: Scale = 1:253,440

Map 2221; 1976: Chapleau-Foleyet : Scale = 1:253,440

Map 2543; 1991: Bedrock Geology of Ontario, East-Central Sheet: Scale 1:1,000,000

Map 2670; 2003: Precambrian Compilation Series, Sault Ste. Marie-Blind River Sheet, Scale = 1:250,000

Preliminary Geologic Map P.248; 1965: Chapleau Sheet, Districts of Sudbury and Manitoulin, Scale = 1:126,720

Preliminary Geologic Map P.423; 1967: Highway 101, Chapleau to Wawa, Scale = 1:63,360

Preliminary Geologic Map P.640, 1971: Wawa Sheet, Scale = 1:126,720

Preliminary Geological Map P.674, 1971: Operation Chapleau, Chapleau Sheet, Scale 1:126,720

*__Note:__ The Preliminary series tend to be very detailed but are usually (not always) in black and white. The regular Map series are usually more general and are usually in color.*